This monograph introduces modern developments on the bound state problem in Schrödinger potential theory and its applications in particle physics.

The Schrödinger equation provides a framework for dealing with energy levels of N-body systems. It was a cornerstone of the quantum revolution in physics of the 1920s but re-emerged in the 1980s as a powerful tool in the study of spectra and decay properties of mesons and baryons. This book begins with a detailed study of two-body problems, including discussion of general properties, level ordering problems, energy-level spacing and decay properties. Following chapters treat relativistic generalizations, and the inverse problem. Finally, three-body problems and N-body problems are dealt with. Applications in particle and atomic physics are considered, including quarkonium spectroscopy. The emphasis throughout is on showing how the theory can be tested by experiment. Many references are provided.

The book will be of interest to theoretical as well as experimental particle and atomic physicists.

CAMBRIDGE MONOGRAPHS ON PARTICLE PHYSICS, NUCLEAR
PHYSICS AND COSMOLOGY: 6

General Editors: T. Ericson, P. V. Landshoff

PARTICLE PHYSICS AND THE SCHRÖDINGER EQUATION

CAMBRIDGE MONOGRAPHS ON PARTICLE PHYSICS, NUCLEAR PHYSICS AND COSMOLOGY

Particle Physics and the Schrödinger Equation

HARALD GROSSE

Institute of Theoretical Physics, University of Vienna

ANDRÉ MARTIN

Theoretical Physics Division, CERN

CAMBRIDGE
UNIVERSITY PRESS

CAMBRIDGE UNIVERSITY PRESS
Cambridge, New York, Melbourne, Madrid, Cape Town, Singapore, São Paulo

Cambridge University Press
The Edinburgh Building, Cambridge CB2 2RU, UK

Published in the United States of America by Cambridge University Press, New York

www.cambridge.org
Information on this title: www.cambridge.org/9780521392259

First published 1997
This digitally printed first paperback version 2005

A catalogue record for this publication is available from the British Library

Library of Congress Cataloguing in Publication data
Grosse, Harald, 1944–
Particle physics and the Schrödinger equation / Harald Grosse, André Martin.
p.   cm. – (Cambridge monographs on particle physics, nuclear physics, and cosmology ; 6)
Includes bibliographical references and index.
ISBN 0-521-39225-X
1. Schrödinger equation.   2. Particles (Nuclear physics)–Mathematics.   3. Two-body problem.
I. Martin, André, Professeur.   II. Title.   III. Series
QC793.3.W3G76   1996
530.1′4–dc20   96-13370   CIP

ISBN-13 978-0-521-39225-9 hardback
ISBN-10 0-521-39225-X hardback

ISBN-13 978-0-521-01778-7 paperback
ISBN-10 0-521-01778-5 paperback

To Heidi and Schu

Contents

Preface

Until 1975 the Schrödinger equation had rather little to do with modern particle physics, with a few exceptions. After November 1974, when it was understood that the J/ψ was made of heavy quark–antiquark pairs, there was a renewed interest in potential models of hadrons, which continued with the discovery of the b quark in 1977. The parallel with positronium was obvious; this is the origin of the neologism "quarkonium". However, in contrast to positronium, which is dominated by the Coulomb potential, the potential between quarks was not known and outside explicit numerical calculations with specific models, there was a definite need for new theoretical tools to study the energy levels, partial widths, radiative transitions, etc. for large classes of potentials. This led to the discovery of a large number of completely new rigorous results on the Schrödinger equation which are interesting not only for the qualitative understanding of quarkonium and more generally hadrons but also in themselves and which can be in turn applied to other fields such as atomic physics. All this material is scattered in various physics journals, except for the *Physics Reports* by Quigg and Rosner on the one hand and by the present authors on the other hand, which are partly obsolete, and the review by one of us (A.M.) in the proceedings of the 1986 Schladming "Internationale Universitätswochen für Kernphysik", to which we will refer later. There was a clear need to collect the most important exact results and present them in an orderly way. This is what we are trying to do in the present book, or least up to a certain cut-off date, since new theorems and new applications continue to appear. This date may look rather far away since it is the beginning of 1995; for instance, the results of J.M. Richard and one of us (A.M.) on the Ω_c particle are not included.

There are two focuses of the book. On the one hand we have rigorous theorems. On the other hand, we have applications to atomic and particle physics which were spectacularly successful, but there is absolutely no attempt to justify at a fundamental level the use of potential models in hadron physics because we feel that its main justification is its success. In addition we felt that we could not avoid presenting a short review of more classical problems like the counting of bound states in potentials, where progress has been made in the last 20 years.

This book does not contain all the material collected in the reviews we mentioned. For instance, the behaviour of the energy levels for large quantum numbers is not reproduced (see the review of Quigg and Rosner and the work of Fulton, Feldman and Devoto both quoted later). The reader will certainly notice, from chapter to chapter, differences in style. However, this book has the merit of being the only one making it possible for a newcomer to become acquainted with the whole subject. Another of its merit is that it does not need any preliminary sophisticated mathematical knowledge. All that is required in most of the book is to know what a second-order differential equation is.

We must warn the reader of the fact that, contrary to common usage, theorems are not numbered separately but like equations, on the right-hand side of the page.

We have to thank many people and primarily Peter Landshoff, who asked us to write this book, and kept insisting, as years passed, until we started working seriously. Our wives, Schu and Heidi, also insisted and we are grateful for that.

Many physicists must be thanked for contributing to the book by their work or by direct help. These are in alphabetical order: B. Baumgartner, M.A.B. Bég J.S. Bell, R. Benguria, R. Bertlmann, Ph. Blanchard, K. Chadan, A.K. Common, T. Fulton, V. Glaser, A. Khare, J.D. Jackson, R. Jost, H. Lipkin, J.J. Loeffel, J. Pasupathy, C. Quigg, T. Regge, J.-M. Richard, J. Rosner, A. De Rújula, A. Salam, J. Stubbe, A. Zichichi.

We would also like to thank Isabelle Canon, Arlette Coudert, Michèle Jouhet, Susan Leech-O'Neale, from the CERN typing pool, for their excellent work in preparing the manuscript in spite of the poor handwriting of one of us (A.M.).

Vienna and Geneva   H. Grosse and A. Martin

1
Overview

1.1 Historical and phenomenological aspects

The Schrödinger equation was invented at a time when electrons, protons and neutrons were considered to be the elementary particles. It was extremely successful in what is now called atomic and molecular physics, and it has been applied with great success to baryons and mesons, especially those made of heavy quark–antiquark pairs.

While before World War II approximation methods were developed in a heuristic way, it is only during the post-war period that rigorous results on the energy levels and the wave functions have been obtained and these approximation methods justified. Impressive global results, such as the proof of the 'stability of matter', were obtained as well as the properties of the two-body Hamiltonians including bounds on the number of bound states. The discovery of quarkonium led to a closer examination of the problem of the order of energy levels from a rigorous point of view, and a comparison of that order with what happens in cases of accidental degeneracy such as the Coulomb and harmonic oscillator potentials. Comparison of these cases also leads to interesting results on purely angular excitations of two-body systems.

Who among us has not written the words 'Schrödinger equation' or 'Schrödinger function' countless times? The next generation will probably do the same, and keep his name alive.

Max Born

Born's prediction turned out to be true, and will remain true for atomic and molecular physics, and — as we shall see — even for particle physics.

When Schrödinger found his equation, after abandoning the relativistic version (the so-called Klein–Gordon equation) because it did not agree

1

with experiments, there was no distinction between atomic, nuclear and particle physics. The wonderful property of the Schrödinger equation is that it can be generalized to many-particle systems and, when combined with the Pauli principle, allows one to calculate, any atom, any molecule, any crystal, whatever their size — at least in principle. The Dirac equation, as beautiful as it may be, is a one-particle equation, and any attempt to generalize it to N-particle systems will have severe limitations and may lead to contradictions if pushed too far, unless one accepts working in the broader framework of quantum field theory.

Because of the capacity of the Schrödinger equation for treating N-body systems it is not astonishing that in the period before World War II all sorts of approximation methods were developed and used, such as the Thomas–Fermi approximation, the Hartree and Hartree–Fock approximations, and the Born–Oppenheimer approximation.

However, except for the fact that it was known that variational trial functions gave upper bounds to the ground-state energies of a system (together with the less well-known min-max principle, which allows one to get an upper bound for the n-th energy level of a system), there was no serious effort to make rigorous studies of the Schrödinger equation. Largely under the impulsion of Heisenberg, simple molecules and atoms were calculated, making chemistry, at least in simple cases, a branch of physics. Also, as was pointed out by Gamow, the Schrödinger equation could be applied to nuclei, which were shown by Rosenblum using an $\propto$ spectrometer to have discrete energy levels.

It was not until after World War II that systematic studies of the rigorous properties of the Schrödinger equation were undertaken. In the 1950s Jost [1], Jost and Pais [2], Bargmann [3], and Schwinger [4] and many others obtained beautiful results on the two-body Schrödinger equation. Then N-body systems were studied, and we shall single out the most remarkable success, namely the proof of the 'stability of matter', first given by Dyson and Lenard [5] and then simplified and considerably improved in a quantitative way by Lieb and Thirring [6], and which is still subject to further study [7]. 'Stability of matter' would be better called the extensive character of the energy and volume of matter: i.e., the fact that NZ electrons and N nuclei of charge Z have a binding energy and occupy a volume proportional to N. Other systems whose behaviour has been clarified in this period are those of particles in pure gravitational interaction [8, 9]. These latter systems do not exhibit the above-mentioned 'stability'; the absolute value of binding energy grows like a higher power of N.

In particle physics, during the 1960s, it seemed that the Schrödinger equation was becoming obsolete, except perhaps in calculating the energy levels of muonic or pionic atoms, or the medium-energy nucleon–nucleon scattering amplitude from a field theoretical potential [10]. It was hoped that elementary particle masses could be obtained from the bootstrap mechanism [11], or with limited but spectacular success from symmetries [12].

When the quark model was first formulated very few physicists considered quarks as particles and tried to calculate the hadron spectrum from them. Among those who did we could mention Dalitz [13], almost 'preaching in the desert' at the Oxford conference in 1965, and Gerasimov [14]. The situation changed drastically after the '1974 October Revolution'. As soon as the J/ψ [15, 16] and the ψ' [17] had been discovered, the interpretation of these states as charm–anticharm bound states was universally accepted and potential models using the Schrödinger equation were proposed [18, 19].

In fact, the whole hadron spectroscopy was reconsidered in the framework of the quark model and QCD in the crucial paper of De Rujula, Georgi and Glashow [20], and the independent papers of Zeldovitch and Sakharov, Sakharov [21], and Federman, Rubinstein and Talmi [22]. Impressive fits of baryon spectra (including those containing light quarks) were obtained, in particular by Stanley and Robson [23, 24], Karl and Isgur [25], Richard and Taxil [26], Ono and Schöberl [27], and Basdevant and Boukraa [28].

We would like to return now to the case of quarkonium — i.e., mesons made of a heavy quark–antiquark pair. By heavy quark, we mean the c and b quarks of effective masses $\sim$ 1.8 and 5 GeV, and also the strange quark, for which the effective mass turns out to be 0.5 GeV. The strange quark occupies a borderline position and can be considered either as heavy, as it is here, or light, as in SU_3 flavour symmetry. In this list one would like to include the top quark, which is certainly heavier than 131 GeV, from the DO experiment [29]. From fits of experimental results by the standard model, including, for instance, masses and widths of the W and Z^0 particles and their partial decays as well as low-energy neutrino experiments (with a standard Higgs), the top quark was predicted to have a mass larger than 150 GeV [30]. Since its mass is heavier than 110 GeV, the notion of toponium becomes doubtful, because the width due to single quark decay, $t \to b + w$, exceeds the spacing between the 1S and the 2S states [31]. In fact the existence of the top quark, is now established with a mass of 175 ± 9 GeV [32].

Figures 1.1 and 1.2 give a summary of the experimental situation for the $c\bar{c}$ (J/ψ etc.) and $b\bar{b}$ (Υ etc.) bound states, respectively.

There are, of course, many potential models used to describe the $c\bar{c}$ and $b\bar{b}$ spectra. The first was a potential

$$V = -\frac{a}{r} + br, \qquad (1.1)$$

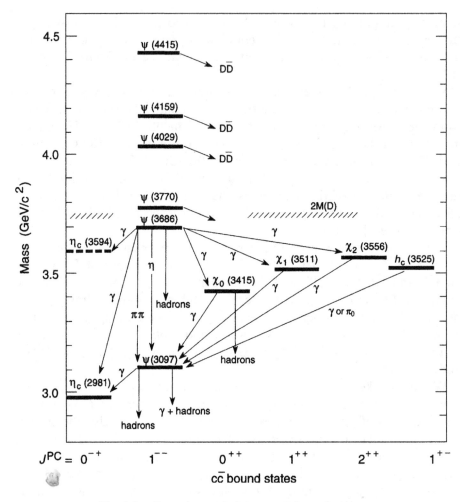

Fig. 1.1. Experimental data on $c\bar{c}$ bound states.

in which the first term represents a one-gluon exchange, analogous to a one-photon exchange, and the second, confinement by a kind of string.

We shall restrict ourselves to two extreme cases of fits. The first, by Buchmüller *et al.* [33], is a QCD-inspired potential in which asymptotic freedom is taken into account in the short-distance part of the potential. The second is a purely phenomenological fit [34] that one of us (A.M.) made with the central potential

$$V = A + Br^{\alpha}. \tag{1.2}$$

Figure 1.3 represents the excitation energies of the $c\bar{c}$ and $b\bar{b}$ systems. The full lines represent the experimental results (for the triplet P states we give only the spin-averaged energies). The dashed lines represent the

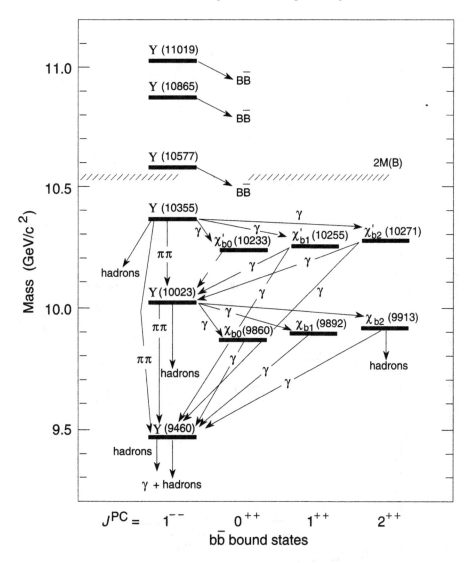

Fig. 1.2. Experimental data on $b\bar{b}$ bound states.

Buchmüller result and the dotted lines result from the potential of Eq. (1.2), to which a zero-range spin–spin interaction $C\delta^3(x)(\vec{\sigma}_1 \cdot \vec{\sigma}_2)/m_1 m_2$ has been added, where m_1 and m_2 are the quark masses, and C was adjusted to the $J/\psi - \eta_c$ separation. The central potential is given by Ref. [34]:

$$V = -8.064 + 6.870 r^{0.1}, \qquad (1.3)$$

where the units are powers of GeV, and quark masses $m_b = 5.174$, $m_c = 1.8$ and, as we shall see, $m_s = 0.518$. The smallness of the exponent, $\alpha = 0.1$, means that we are very close to a situation in which the spacing of energy

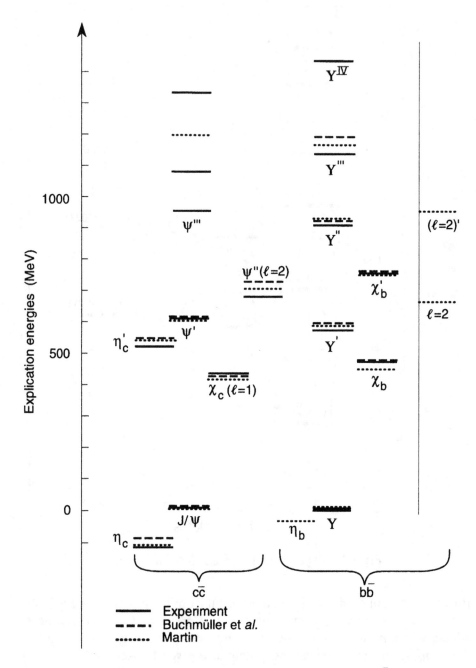

Fig. 1.3. Comparison of the excitation energies of the $c\bar{c}$ and $b\bar{b}$ systems with two theoretical models.

Table 1.   Relative leptonic widths.

	Experiment	Buchmüller	Martin
ψ'	0.46 ± 0.6	0.46	0.40
ψ'''	0.16 ± 0.02	0.32	0.25
Υ'	0.44	0.44	0.51
Υ''	0.33	0.32	0.35
Υ'''	0.20	0.26	0.27

levels is independent of the mass of the quarks and the case for a purely logarithmic potential.

Table 1 represents the relative leptonic widths — i.e., the ratios of the leptonic width of a given $\ell = 0$ state to the leptonic width of the ground state. 'Theory' uses the so-called Van Royen-Weisskopf formula.

We see that both the fits are excellent. The QCD-inspired fit reproduces somewhat better the low-energy states, in particular the separation between the $\ell = 0$ and $\ell = 1$ states for the $b\bar{b}$ system. This is presumably due to the fact that the QCD-inspired potential has a correct short-range behaviour while the phenomenological potential has not (there is a discrepancy of 40 MeV, which would have been considered negligible before 1974, but with the new standards of accuracy in hadron spectroscopy can no longer be disregarded). On the other hand, the phenomenological potential gives a better fit for higher excitations, those close to the dissociation threshold into meson pairs $D\bar{D}$, $B\bar{B}$. This may be due to the fact that the optimal $\alpha = 0.1$ takes into account the lowering of the energies of confined channels $c\bar{c}$, $b\bar{b}$, due to their coupling to open channels.

In the list of parameters of the phenomenological potential, we have already indicated the strange-quark effective mass $m_s = 0.518$ GeV. This is because, following the suggestion of Gell-Mann, we have pushed, the phenomenological model beyond its limit of validity! Remarkably, one gets a lot of successful predictions. $M_\phi = 1.020$ GeV is an input but $M_{\phi'} = 1.634$ GeV agrees with experiment (1.650 GeV). At the request of De Rujula the masses of the $c\bar{s}$ states have also been calculated. One gets

$$M_{D_s} = 1.99 \ (\text{exp } 1.97, \text{previously } 2.01)$$

$$M_{D_s^*} = 2.11 \ (\text{exp } 2.11) \,,$$

and in 1989 Argus [35] observed what is presumably a $\ell = 1$ $c\bar{s}$ state, which could be $J^P = 1^+$ or 2^+ of mass 2.536 GeV. The spin-averaged mass of such a state was calculated, without changing any parameter of the model, and

$$M_{D_s^{**}} (\ell = 1) = 2.532$$

was obtained [36]. One could conclude that the state observed by Argus is no more than 30 MeV away from the centre of gravity.

More recently, a B_s meson was observed, both at LEP and at Fermilab. The least square fit to the mass turns out to be 5369 ± 4 MeV [37], while the theoretical prediction of the model is 5354–5374 MeV [38]. It is impossible not to be impressed by the success of these potential models. But why are they successful? The fact that the various potentials work is understood: different potentials agree with each other in the relevant range of distances, from, say, 0.1 fermi to 1 fermi. However, relativistic effects are not small; for the $c\bar{c}$ system, v^2/c^2 is calculated *a posteriori* to be of the order of 1/4.

The sole, partial explanation we have to propose is that the potential is simply an effective potential associated with an effective Schrödinger equation. For instance, one can expand $\sqrt{p^2 + m^2}$, the relativistic kinetic + mass energy, around the average $\langle p^2 \rangle$ instead of around zero. For a purely logarithmic potential the average kinetic energy is independent of the excitation, and it happens that the potential is not far from being logarithmic. Anyway, we must take the pragmatic attitude that potential models work and try to push the consequences as far as we can.

Concerning baryons, we shall be more brief. Baryons made purely of heavy quarks, such as bbb and ccc, have not yet been found, though they must exist. Bjorken [39] advocates the study of ccc, which possesses remarkable properties: it is stable against strong interactions and has a lifetime which is a fraction of 10^{-13} seconds. Its lowest excitations are also stable or almost stable. If one accepts that the quark–quark potential inside a baryon is given by [40]

$$V_{QQ} = \frac{1}{2} V_{Q\bar{Q}}, \tag{1.4}$$

one can calculate all the properties of ccc from a successful $c\bar{c}$ potential. Bjorken thinks that such a state can be produced at a rate not-too-small to be observed.

In the meantime we should remember that the strange quark can be regarded as heavy. J.-M. Richard, using the fit (1.3) of quarkonium and rule (1.4), has obtained a mass for the Ω^- baryon sss of 1.665 GeV [41] while experiment [42] gives 1.673 GeV.

For baryons made of lighter quarks, following the pioneering work of Dalitz came the articles of De Rujula, Georgi, Glashow [20], Zeldovitch and Sakharov, Sakharov alone [21], and Federman, Rubinstein and Talmi [22]. In these works, the central potential is taken to be zero or constant

Table 2.   Masses for $V = A + B\tau^{0.1}$.

	Theory	Experiment
N	input	0.939
Δ	input	1.232
$\Lambda^{\bullet}$	1.111	1.115
Σ	1.176	1.193
Ξ	1.304	1.318
Σ^{*}	1.392	1.383
Ξ^{*}	1.538	1.533
Ω^{-}	input	1.672
Λ_c	input	2.282
Σ_c	2.443	2.450
Σ_c^{*}	2.542	
$\Xi_c = A$	2.457	2.460
S	2.558	
S^{*}	2.663	

— i.e., incorporated in the quark masses and the dominant feature is given by the spin–spin forces 'derived' from QCD, which lead to remarkable results, in particular the first explanation of the $\Sigma - \Lambda$ mass difference. In this approach, which is zero order in the central potential, the calculation of excited states is excluded.

The next step is to add a soft central potential and try to solve accurately the three-body Schrödinger equation. This has been done by many people. Stanley and Robson [24] were among the first, and Karl, Isgur, Capstick and collaborators [25, 43] were among the most systematic.

Here we would like to limit ourselves to the study of ground states, which has been done, for instance, by Ono and Schöberl [27] and Richard and Taxil [26]. For example, we would like to show, in Table 2, the results of Richard and Taxil with a potential $V = A + Br^{0.1}$ and a spin–spin Hamiltonian

$$C \frac{\vec{\sigma}_i \vec{\sigma}_j}{m_i m_j} \delta (\vec{r}_i - \vec{r}_j) \ . \tag{1.5}$$

Although the results are nice, it is not completely obvious whether the rigorous treatment of the central potential does lead to a real improvement. To demonstrate this, we have taken some ratios, which in the De Rujula, Georgi and Glashow model [20] have simple values, and have shown in Table 3 a comparison of the calculated and experimental values.

Perhaps it is worth noting that the equal-spacing rule of the SU_3 flavour

Table 3.   Ratio of mass differences including the Gell-Man–Okubo predictions (G.M.O) compared to experiment.

	De Rujula Georgi Glashow Sakharov Zeldovitch Federman Rubinstein Talmi	Richard Taxil	Experiment
$(M_{\Xi^*} - M_\Xi)/(M_{\Sigma^*} - M_\Sigma)$	1	1.08	1.12
$(2M_{\Sigma^*} + M_\Sigma - 3M_\Lambda)/2(M_\Delta - M_N)$	1	1.07	1.05
$(3M_\Lambda + M_\Sigma)/(2M_N + 2M_\Xi)$ GMO OCTET	1	1.005	1.005
$(M_{\Sigma^*} - M_\Lambda)/(M_{\Xi^*} - M_{\Sigma^*})$ GMO	1	1.10	1.03
$(M_{\Xi^*} - M_{\Sigma^*})/(M_{\Omega^-} - M_{\Xi^*})$ DECUPLET	1	1.09	1.08

decuplet, which was the triumph of Gell-Mann, enters here in a rather unusual way. Naturally, if the spin–spin forces and the central potential are neglected, the equal-spacing rule is absolutely normal, since the mass of the state is obtained by merely adding the quark masses: therefore the mass is a <u>linear</u> function of strangeness.

However, Richard and Taxil [44] discovered by numerical experiments that if one takes a 'reasonable', flavour-independent, two-body central potential the masses of the decuplet are <u>concave</u> functions of strangeness. In other words,

$$M(ddd) + M(dss) < 2M(dds). \tag{1.6}$$

1.2   Rigorous results

This is perhaps a good point to turn to the main part of the book, which concerns rigorous results on the Schrödinger equation stimulated by potential models.

Returning to inequality (1.6), we can state the theorem obtained by Lieb [45]:

M is a concave function of the strange quark mass if the two-body potential V(r) is such that

$$V' > 0, \ V'' < 0, \ V''' > 0.$$

This is the case, for instance, for

$$V = -\frac{a}{r} + br, V = r^{0.1} \text{ etc.}$$

(However the property is <u>not true</u> for power potentials with large exponents, such as, $V = r^5$ [44].)

Now let us return to the linearity in the decuplet: this comes from a cancellation between the effect of the central potential and the spin-dependent potential.

What we have just seen is only one example: the calculation of the energy levels of the decuplet leading to the discovery of the property of concavity with respect to the strange-quark mass for a certain class of potentials.

Now we shall concentrate on two-body systems. We ask the reader to return to Figures 1.1 and 1.2, the $c\bar{c}$ and $b\bar{b}$ spectra. A remarkable property is that the average P state ($\ell = 1$) mass is <u>below</u> the first ($\ell = 0$) excitation. This property is satisfied by all existing models, in particular the initial model [18, 19] which made this statement <u>before</u> the discovery of the P state (at the Dijon congress of the Societé Française de Physique, Gottfried said that if the P states failed to be at the right place their model would be finished. An experimentalist from SLAC who knew that there was evidence for these P states kept silent!). One of the authors (A.M.) was asked by Bèg, during a visit at Rockefeller University, if this prediction was typical of the model and could be changed by modifying the potential. The problem, therefore, was to find a simple criterion to decide the order of levels in a given potential. Although some preliminary results were obtained in 1977 by the present authors, it was not until 1984 that the situation was completely clarified.

We denote by $E(n, \ell)$ the energy of the state with an angular momentum ℓ and radial wave function with n nodes. For a general potential we have only the restrictions

$$E(n + 1, \ell) > E(n, \ell), \tag{1.7}$$

$$E(n, \ell + 1) > E(n, \ell), \tag{1.8}$$

which follow respectively from Sturm–Liouville theory and from the positivity of the centrifugal term. What we want is more than that!

There are two potentials for which the solutions of the Schrödinger equation are known for all n and ℓ and which exhibit 'accidental' degeneracies, the Coulomb potential and the harmonic oscillator potential. In fact, one can go from one to the other by a change of variables. The Coulomb potential can be characterized by

$$r^2 \Delta V(r) = 0 , \qquad (1.9)$$

i.e., the Laplacian of the potential is zero outside the origin, while the harmonic oscillator potential satisfies

$$\frac{d}{dr}\frac{1}{r}\frac{dV}{dr} = 0 , \qquad (1.10)$$

i.e. it is linear in r^2.

Now, in the case of quarkonium, what can we say about the potential? First of all we have 'asymptotic freedom'. 'Strong' asymptotic freedom is when the <u>force</u> between quarks is $-\alpha(r)/r^2$, with

$$\frac{d}{dr}\alpha(r) > 0 , \qquad (1.11)$$

which is equivalent to

$$r^2 \Delta V(r) > 0 . \qquad (1.12)$$

On the other hand, according to Seiler [46], lattice QCD implies that V is <u>increasing</u> and <u>concave</u>, which in turn implies

$$\frac{d}{dr}\frac{1}{r}\frac{dV}{dr} < 0 . \qquad (1.13)$$

The following are the two main theorems [47] which are relevant to this situation (and, as we shall see, to other situations).

Theorem:

$$E(n,\ell) \gtrless E(n-1,\ell+1)$$

$$\text{if} \quad r^2 \Delta V(r) \gtrless 0 \ \ \forall r > 0 . \qquad (1.14)$$

Theorem:

$$E(n,\ell) \gtrless E(n-1,\ell+2)$$

$$\text{if} \quad \frac{d}{dr}\frac{1}{r}\frac{dV}{dr} \gtrless 0 \ \ \forall r > 0 , \qquad (1.15)$$

i.e., if V is convex or concave in r^2.

There are, however, other important results from which we choose the following:

Theorem: [48]

$$E(0,\ell) \text{ is convex (concave) in } \ell$$

$$\text{if} \quad V \text{ is convex (concave) in } r^2 . \tag{1.16}$$

$E(0,\ell)$ is what used to be called the 'leading Regge trajectory' in the 1960s!

Theorem: [49]

(a) The spacing between $\ell = 0$ energy levels increases (decreases) with n:

$$E(n+2,0) - E(n+1,0) \gtrless E(n+1,0) - E(n,0)$$

$$\text{if} \quad V = r^2 + \lambda v, \quad \lambda \text{ small and } \frac{d}{dr} r^5 \frac{d}{dr} \frac{1}{r} \frac{dv}{dr} \gtrless 0 \,\forall r \,; \tag{1.17}$$

(b) it <u>decreases</u> if $V'' < 0$ (the proof of this is not yet complete, so that it is really only a conjecture which we believe with a 99.9% probability).

Theorem: [50]
If $r^2 \Delta V \gtrless 0$ and

$$\frac{d^2 E (0,\ell)}{d\ell^2} + \frac{3}{\ell+1} \frac{dE (0,\ell)}{d\ell} \gtrless 0 , \tag{1.18}$$

this implies

$$\frac{E (0,\ell+1) - E (0,\ell)}{E (0,\ell) - E (0,\ell-1)} \gtrless \frac{E_c (0,\ell+1) - E_c (0,\ell)}{E_c (0,\ell) - E_c (0,\ell-1)} , \tag{1.19}$$

where E_c is the Coulomb energy

$$E_c = -\text{const}(\ell + 1)^{-2} .$$

Figure 1.4 illustrates Theorem (1.14). The unbroken lines represent the pure Coulomb case, the dashed lines the case $r^2 \Delta V > 0$, and the dotted lines the case $r^2 \Delta V < 0$.

Now the application to quarkonium is obvious. We see this very clearly in Figures 1.1 and 1.2, and also that it corresponds to $r^2 \Delta V \geq 0$. There are, however, other applications.

The first of these which we shall consider is to muonic atoms, where the size of the nucleus cannot be neglected with respect to the Bohr

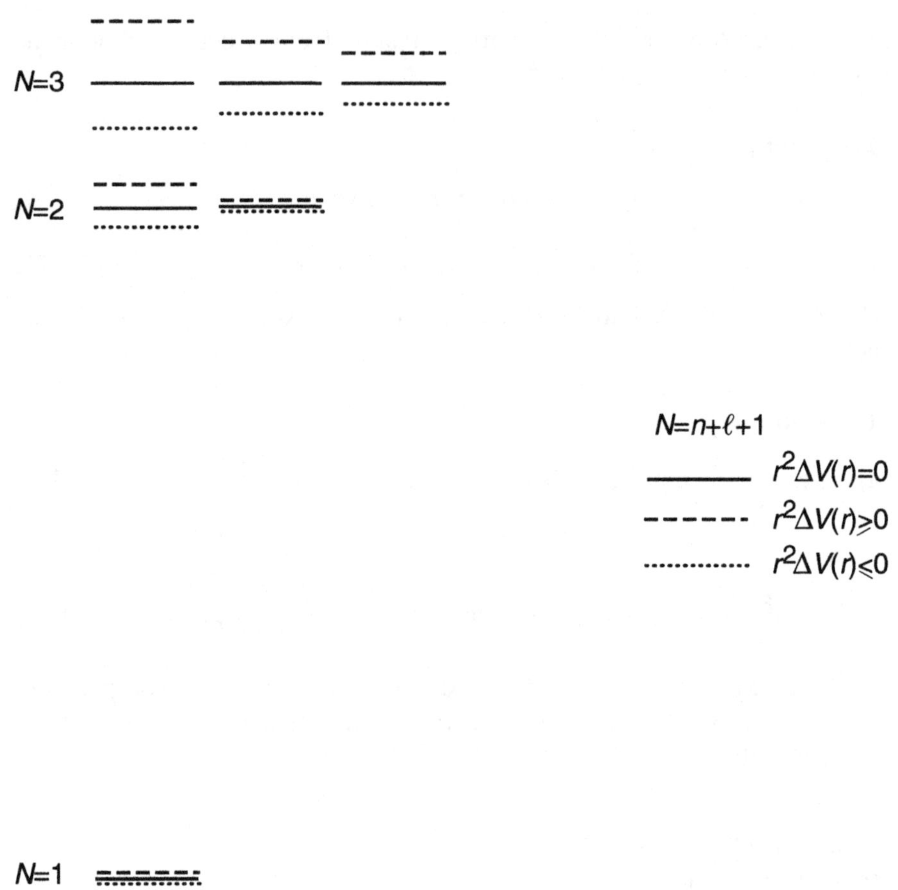

$N=3$

$N=2$

$N=n+\ell+1$

$r^2 \Delta V(r)=0$

$r^2 \Delta V(r) \geq 0$

$r^2 \Delta V(r) \leq 0$

$N=1$

Fig. 1.4. Illustration of the order of levels for potentials with zero, positive or negative Laplacian.

orbit. Since the nucleus has a positive charge distribution, the potential it produces, by Gauss's law, has a positive Laplacian. In the tables by Engfer *et al.* [51] one finds abundant data on $\mu^{-\,138}$Ba atoms.

In particular, in spectroscopic notation, one finds for the $N = 2$ levels

$$2s_{1/2} - 2p_{1/2} = 405.41 \text{ keV} ,$$

which is <u>positive</u> and which would be zero for a point-like nucleus (even for the Dirac equation). Similarly, for $N = 3$ one has

$$3d_{3/2} - 2p_{1/2} = 1283.22 \text{ keV} ,$$

while

$$3p_{3/2} - 2p_{1/2} = 3p_{3/2} - 2s_{1/2} + 2s_{1/2} - 2p_{1/2} = 1291.91 \text{ keV} ,$$

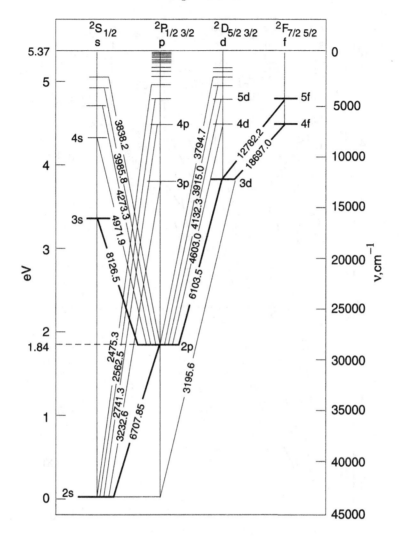

Fig. 1.5. Spectrum of the lithium atom.

which means that

$$3p_{3/2} \text{ is above } 3d_{3/2}.$$

For $N = 4$ a violation of 4 keV is found, but this is compatible with relativistic corrections.

Another application is to alkaline atoms. In these, the outer electron is subjected to the potential produced by a point-like nucleus and by the negatively charged electron cloud, so that in the Hartree approximation we have $r^2 \Delta V < 0$. Figures 1.5 and 1.6 illustrate this situation for lithium and sodium, respectively [52].

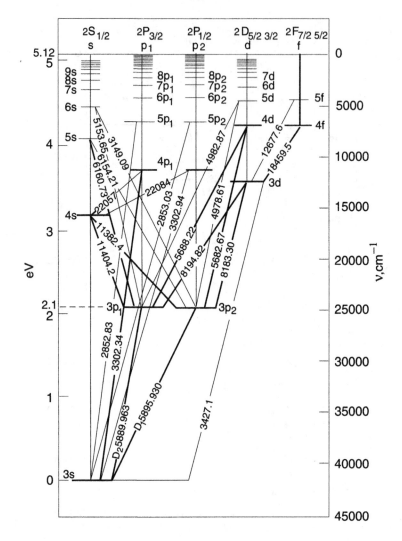

Fig. 1.6. Spectrum of the sodium atom.

Now we turn to Theorem (1.15). Figure 1.7 shows the energy level diagram of the harmonic oscillator (unbroken line), of a potential with $(d/dr)(1/r)(dV/dr) > 0$ (dashed line), and of a potential with $(d/dr)(1/r)(dV/dr) < 0$ (dotted line).

Of course, the $\ell = 2$ state of the $c\bar{c}$ system, called ψ'', satisfies this theorem:

$$M_{\psi''} = 3.77 \text{ GeV, while } M_{\psi'} = 3.68 \text{ GeV .}$$

Next, we would like to illustrate Theorem (1.19) [50], which concerns

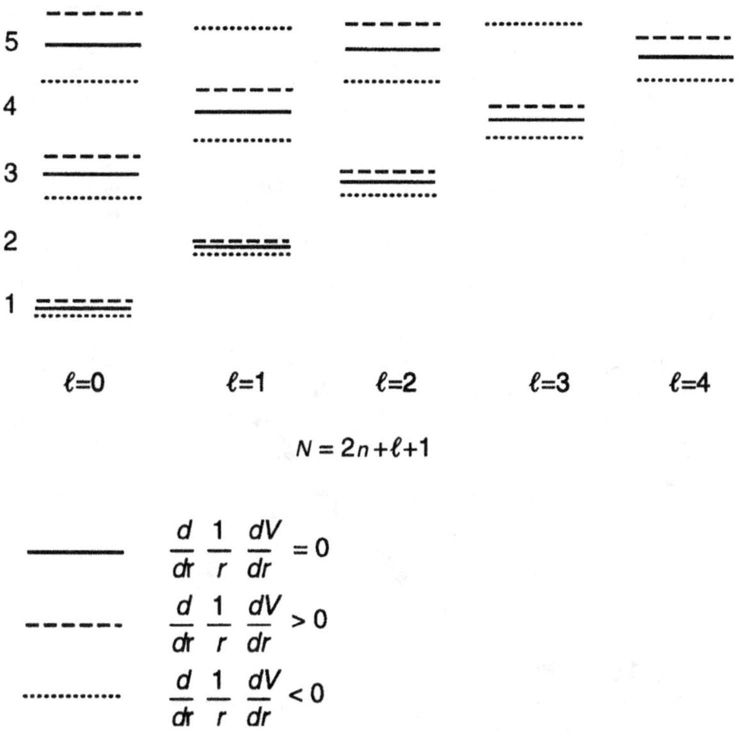

$$N = 2n + \ell + 1$$

————	$\dfrac{d}{dr}\dfrac{1}{r}\dfrac{dV}{dr} = 0$
— — — —	$\dfrac{d}{dr}\dfrac{1}{r}\dfrac{dV}{dr} > 0$
··········	$\dfrac{d}{dr}\dfrac{1}{r}\dfrac{dV}{dr} < 0$

Fig. 1.7. Illustration of the order of levels for potentials linear, convex or concave in r^2.

purely <u>angular</u> excitations. If $r^2 \Delta V \gtrless 0$

$$\frac{E_{3d} - E_{2p}}{E_{2p} - E_{1s}} \gtrless 0.185... , \qquad (1.20)$$

$$\frac{E_{4f} - E_{3d}}{E_{3d} - E_{2p}} \gtrless 0.35 , \qquad (1.21)$$

$$\frac{E_{5g} - E_{4f}}{E_{4f} - E_{3d}} \gtrless 0.463... . \qquad (1.22)$$

Inequality (1.20) with $>$ is, of course, satisfied by the $c\bar{c}$ system

$$\frac{M_{\psi''} - M_{\chi c}}{M_{\chi c} - M_{J/\psi}} = \frac{3.77 - 3.51}{3.51 - 3.10} = 0.634 \gg 0.185,$$

but this is not very exciting. It is, we believe, in muonic atoms [51] that we find the most spectacular illustration.

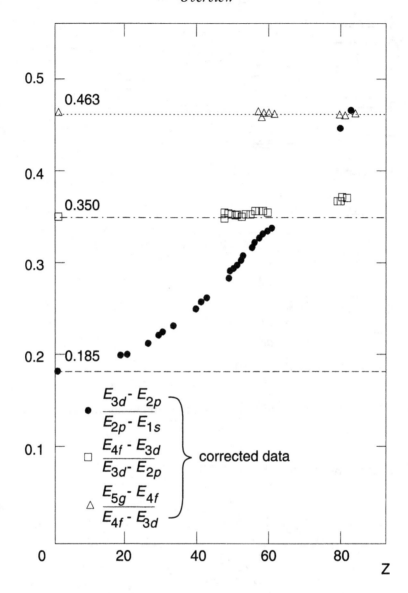

Fig. 1.8. Ratios of spacings between angular excitations of muonic atoms as a function of the charge of the nucleus.

In Figure 1.8 we have represented the ratios (1.20), (1.21) and (1.22) as a function of the charge Z of the nucleus. Relativistic effects have been eliminated by a procedure that we shall not describe here. We see that the first two ratios deviate from the Coulomb value very clearly as Z increases, while the last, which is insensitive to the non-zero size of the nucleus, remains constant.

Table 4. Ratios of spacing of angular excitations for the lithium and sodium sequences.

		Lithium sequence		Sodium sequence		
		$\dfrac{4f-3d}{3d-2p}$	$\dfrac{5g-4f}{4f-3d}$		$\dfrac{5g-4f}{4f-3d}$	$\dfrac{6h-5g}{5g-4f}$
Coulomb		0.35	0.4638	Coulomb	0.4638	0.5432
Li	I	0.326				
Be	II	0.324	0.462			
B	III	0.326	0.462			
C	IV	0.329	0.462			
N	V	0.331	0.462			
O	VI	0.331				
F	VII	0.334	0.462			
Ne	VIII	0.336				
Na	IX			Na	I	0.456	0.5426
Mg	X	0.338		Mg	II	0.444	0.5419
Al	XI	0.339		Al	IV	0.432	0.5413
Si	XII	0.339		Si	IV	0.423	0.5409
P	XIII	0.339		P	V	0.418	0.5406
S	XIV	0.340		S	VI	0.416	

An illustration going in the opposite direction is obtained by looking at the spectrum of lithium in Figure 1.5. We find

$$\frac{E_{4f} - E_{3d}}{E_{3d} - E_{2p}} = 0.326 < 0.35$$

as we should.

Table 4 shows some ratios of differences of energy levels for the lithium and sodium sequences, using the Bashkin and Stoner tables [53]. There are also inequalities in the ionization energies, which can be obtained and used as a test for the claimed accuracy of the experimental data.

One can even go further and incorporate the spin of the muon or of the outer electron and get interesting inequalities in the fine structure splittings of the purely angular excitations treated in a semirelativistic way. An application to muonic atoms shows that for Z less than 40 the actual values and the bounds differ by less than 20%. In the case of alkaline atoms the inequalities are marginally satisfied by the lithium sequence but fail for the sodium sequence. This is an indication of the failure of the Hartree approximation for these systems and is expected, since the sodium doublet is inverted — i.e., the state with higher J is lower.

Next, we shall make a remark about the P-state splitting of the $c\bar{c}$ singlet as well as the $b\bar{b}$ system. The P states are split by spin–spin, spin-orbit and tensor forces. One of the authors (A.M.), together with Stubbe [54], has assumed that the splitting is given by the Fermi–Breit Hamiltonian and that the central potential contains only vector-like and scalar-like components. It has then been possible to bound the difference of the mass of the $^1P_1$ state and the weighted average of the triplet P states (S = 1). With this strategy, and using experimental numbers and allowing for some relativistic corrections, the bounds

$$3536 \pm 12 < M(^1P_1) < 3559 \pm 12 \text{ MeV}$$

for the $c\bar{c}$ system and, similarly, the bounds

$$9900.3 \pm 2.8 < M(^1P_1) < 9908.9 \pm 2.8 \text{ MeV},$$

for the $\bar{b}b$ system were obtained. The experimental result for the $c\bar{c}$ $^1P_1$ state is 3526.1 MeV, which is compatible with the bounds and the early indications from the ISR.

Other narrow states still to be observed are the 1D 2^{--} and 1D 2^{-+} of the $c\bar{c}$ system as well as the two complete sets of D states of the $b\bar{b}$ system, for which Kwong and Rosner [55] give predictions based on the 'inverse scattering' method, shown in Figure 1.9.

During 1991 and 1992 relativistic effects were also investigated: i.e., we looked at particles satisfying the Klein–Gordon and Dirac equations. In the Klein–Gordon equation, if V is attractive and $\Delta V \leq 0$ one has [56]

$$E(n+1, \ell) < E(n, \ell+1), \tag{1.23}$$

i.e., the levels are ordered like those of a Schrödinger equation with a potential having a negative Laplacian.

There is a converse theorem for $\Delta V > 0$, but it is more sophisticated: one has to replace ℓ by another 'effective' angular momentum.

Concerning the Dirac equation, one of the authors (H.G.) had already obtained a perturbative result in Ref. [57]. This was that, for perturbations around the Coulomb potential, levels with the same J and different ℓs are such that

$$E(N, J, \ell = J - 1/2) \gtrless E(N, J, \ell = J + 1/2) \tag{1.24}$$

if $\Delta V \gtrless 0$ for all $r > 0$. The principal quantum number in (1.24) is denoted by N. This means that the order of levels holds not only in the Schrödinger case but also in the Dirac case, if one accepts the validity of the result even for larger V. It also means that the Lamb shift effect is equivalent to replacing the source of the Coulomb potential by an extended structure. It remains a challenge to find a non-perturbative version of this result.

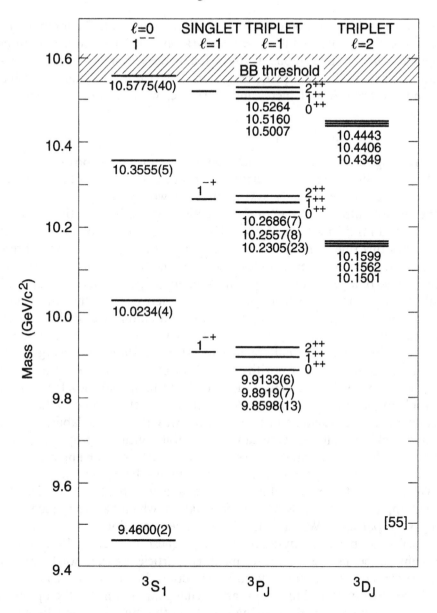

Fig. 1.9. $b\bar{b}$ level diagram according to Kwong and Rosner [55]

On the other hand, we have been able together with Stubbe to compare levels of the Dirac equation for the same N, the same <u>orbital</u> angular momentum and different total angular momenta [58] and have found

$$E\left(N, J = \ell + \frac{1}{2}, \ell\right) > E\left(N, J = \ell - \frac{1}{2}, \ell\right) , \qquad (1.25)$$

if $dV/dr > 0$. If we combine the two results, we see that a Coulomb multiplet is completely ordered if $\Delta V < 0$ and $dV/dr > 0$. The second condition might actually be superfluous, as suggested by the semi-relativistic approximation. Then, in the multiplet, energies increase for fixed J and increasing L, and for fixed L and increasing J. Hence, we have, for instance, for $N = 3$,

$$3S_{1/2} < 3P_{1/2} < 3P_{3/2} < 3D_{3/2} < 3D_{5/2} . \tag{1.26}$$

Future progress on the one-particle Dirac equation seems possible.

There are also a number of relatively recent rigorous results on the three-body and even the N-body system that we would like to mention. We have already introduced the 'rule' connecting two-body potentials inside a baryon and inside a meson, $V_{QQ} = \frac{1}{2} V_{Q\bar{Q}}$ at the phenomenological level. If one takes this rule seriously one can calculate a lower bound for a three-body Hamiltonian in terms of two-body bound states. This is, in fact, a special case of a general technique invented long ago which can be applied with success to any N-particle system with attractive forces, including three quarks and N particles in gravitational interaction. We shall describe refinements of this technique which lead to very accurate lower bounds (i.e., very close to variational upper bounds). For instance, the energy of N gravitating bosons is known with an accuracy of less than 7% for arbitrary N. All these results are obtained at the non-relativistic level, but it is possible to make a connection between semirelativistic and non-relativistic treatments and to demonstrate in a simple way the unavoidability of the Chandrasekhar Collapse. One can also exploit concavity properties, with respect to the inverse of the mass of a particle, to obtain upper bounds on the masses of baryons or mesons containing a heavy quark.

We hope that our contribution will enable the reader to realize the broad application of the Schrödinger equation, which has been uncovered only very partially. Whether the Schrödinger equation will continue to be useful for particle physics is an open question. Before 1977 it was thought to be useless in the elementary particle physics world except for describing the nucleon–nucleon interaction, but that has changed as we have already seen. This might again change, next time in the opposite direction. However, the results obtained under the stimulus of the discovery of heavy-quark systems will remain and may be useful in other areas, such as atomic physics or even condensed matter physics.

At the end of this introduction let us make a few remarks about the literature. The subject started to be examined in the middle of the 1970s and two reviews appeared a little later [59, 60]. Since then, a large amount of new material, both theoretical and experimental, has come into existence. This is partially summarized in reviews by one of us (A.M.) [61–63] and also in Ref. [64].

2

Two-body problems

2.1   General properties

We shall discuss at the beginning of this section a number of general properties of interactions via a potential. The purpose is to explain a few facts in simple terms without entering into too-heavy mathematics. To begin with we quote once and for all the Schrödinger equation for a two-particle system. If the potential depends only on the relative positions, it becomes in the centre-of-mass system

$$\left(-\frac{\hbar^2}{2m}\Delta + V(\vec{x})\right)\psi(\vec{x}) = E\psi(\vec{x}) . \qquad (2.1)$$

We shall mostly deal with a central potential in three dimensions. Only some results of Section 2.7 and Part 3 hold for non-central potentials and some in any dimension. Radial symmetry of the potential allows further simplification by taking $\psi(\vec{x}) = Y_{\ell m}(\Omega)u_{n,\ell}(r)/r$. Here, $Y_{\ell m}$ denotes the ℓ, m-th spherical harmonic function. For the reduced radial wave function $u_{n,\ell}(r)$ Eq. (2.1) becomes

$$\left(-\frac{\hbar^2}{2m}\frac{d^2}{dr^2} + \frac{\ell(\ell+1)\hbar^2}{2mr^2} + V(r)\right)u_{n,\ell}(r) = E(n,\ell)u_{n,\ell}(r) , \qquad (2.2)$$

where m denotes the reduced mass of the two-particle system. We shall often put $\hbar = 2m = 1$ to simplify the presentation. Clearly we have to assume a few facts in order to have a well-defined problem. For stability reasons we would like the Hamiltonian entering the Schrödinger equation to be lower bounded. We are therefore dealing mostly with regular potentials which are not too singular at the origin and which either decrease to zero or are confining at infinity. We shall also require that $|V|^{3/2}$ be locally integrable in three dimensions. This guarantees the

lower boundedness of the Hamiltonian. For the spherically symmetric case
we like to have finiteness of $\int_0^R dr\, r|V(r)|$. For non-confining potentials
the last condition with $R = \infty$ implies finiteness of the number of bound
states. Regular potentials are also given if $V(r)$ is less singular than $-1/4r^2$
at the origin. If $\lim_{r\to 0} r^2 V(r) = 0$, $u_{n,\ell}(r)$ is proportional to $r^{\ell+1}$ near the
origin. An intuitive picture of the behaviour of $u_{n,\ell}(r)$ can be gained very
easily. For simplicity we take $\ell = 0$, $\hbar = 2m = 1$, denote $u_{n,\ell}$ by u, $E(n,\ell)$
by E, and assume that $V(r)$ is monotonously increasing and $V(0)$ is finite.
Equation (2.2) then becomes $u'' = (V - E)u$. We start with $u(r) = r$ for
r small and integrate to infinity, varying the parameter E. We can first
take $E < V(0) \le V(r)$. Then u'' is always positive, and therefore convex
and goes to infinity. The normalization condition $\int_0^\infty dr|u(r)|^2 = 1$ can
never be obtained. Increasing E such that $V(0) < E = V(r_{cl})$, where
r_{cl} denotes the classical turning point, yields two intervals with different
behaviours. For $0 < r < r_{cl}$, u is concave; for $r_{cl} < r < \infty$ it is convex. It is
therefore understandable that at a large enough value of E (the ground-
state energy) $u(r)$ will tend to zero for $r \to \infty$ and becomes normalizable.
In addition, no node will be present. Increasing E still further will first
yield a solution to the differential equation going to $-\infty$, which has
one zero — within $(0, \infty)$ — since the curvature of $u(r)$ changes if it
becomes zero. Increasing E still further will yield an energy eigenvalue
corresponding to a radial excited state (if it exists), which has one zero,
etc. Such simple convexity conditions together with the nodal structure
have been used by one of the authors (H.G.) [65] in order to obtain
a systematic numerical procedure for locating bound-state energies. We
remark clearly that if $V(\infty) = 0$ we obtain scattering solutions for $E > 0$
and possible resonance behaviour for $E = 0$. According to the above
simple arguments (which we shall show analytically later) we may label
u and E by n, the number of nodes u has within $(0, \infty)$, and the angular
momentum quantum number ℓ.

We have assumed $V(r)$ to be 'smooth' so that (2.2) defines a self-adjoint
operator which has only a real spectrum. For a large proportion of this
book, $V(r)$ will be taken to be a confining potential which goes to infinity
for $r \to \infty$ and which has therefore only a discrete spectrum. In some
places we shall deal with a potential going to zero. In that case oscillations
of $V(r)$ could even produce a bound state for positive E. The first, very
simple example is due to Von Neumann: one starts with a candidate for
a bound-state wave function

$$u(r) = \frac{\sin kr}{C + \int_0^r dr'(\sin kr')^2} \equiv \frac{\sin kr}{D} \underset{r\to\infty}{\sim} \frac{2\,\sin kr}{r}, \qquad (2.3)$$

which is square integrable on the half-line $[0,\infty)$. Differentiating (2.3) twice and dividing through $u(r)$ we get

$$\frac{u''}{u} = -k^2 + \left(\frac{2(\sin kr)^4}{D^2} - \frac{4k \cos kr \sin kr}{D}\right), \qquad (2.4)$$

which shows that u corresponds to a bound state at positive energy $E = k^2$ for a potential, which oscillates at infinity:

$$V(r) \simeq -\frac{4k \sin 2kr}{r}. \qquad (2.5)$$

Such pathological cases are easily excluded. One sufficient condition excluding such cases is given if V goes monotonously to zero beyond a certain radius, $r > r_0$. Other sufficient conditions are $\int_R^\infty dr|V'(r)| < \infty$ or $\int_R^\infty dr|V(r)| < \infty$. In all three cases all states with positive energy correspond to scattering states.

Let us point out, however, that there exist 'good' potentials, for which the Hamiltonian is self-adjoint, the spectrum lower-bounded, and the bound states have negative energy, which are oscillating violently at the origin or at infinity as discovered by Chadan [66]. It is the primitive of this potential which must possess regularity properties.

There are a few simple principles which are helpful in locating bound states. The ground-state energy E_1 of the Hamiltonian H, for example, is obtained as the infimum

$$E_1 = \inf_{\psi \in \mathscr{H}} \frac{\langle \psi|H\psi\rangle}{\langle \psi|\psi\rangle}, \qquad (2.6)$$

where $\mathscr{H}$ denotes the Hilbert space of L^2-functions $\psi \neq 0$, and $\langle \cdot|\cdot\rangle$ the scalar product. From (2.6) it follows that any trial function φ will yield an upper bound on $E_1 \leq \langle \varphi|H\varphi\rangle/\langle \varphi|\varphi\rangle$. In addition, we deduce from (2.6) that adding a positive potential (in the sense that all expectation values $\langle \chi|V\chi\rangle$ are positive) can only increase the energy. In order to obtain the n-th excited state one has to take first the maximal eigenvalue within an n-dimensional space $\mathscr{H}_n$ spanned by linear independent trial functions to get an upper bound on E_n [67]

$$E_n = \inf_{\mathscr{H}_n} \max_{\psi \in \mathscr{H}_n} \langle \psi|H\psi\rangle/\langle \psi|\psi\rangle. \qquad (2.7)$$

It follows, that the addition of a positive potential V increases all eigenvalues: we denote by $\phi_1, \ldots, \phi_n$ the first n eigenfunctions of $\bar{H} = H + V$ to eigenvalues $\bar{E}_n$, and by $\mathscr{D}_n$ the linear space spanned by $\{\phi_1, \ldots, \phi_n\}$. Then

$$E_n \leq \max_{\psi \in \mathscr{D}_n} \langle \psi|H\psi\rangle/\langle \psi|\psi\rangle \leq \max_{\psi \in \mathscr{D}_n} \langle \psi|\bar{H}\psi\rangle/\langle \psi|\psi\rangle = \bar{E}_n. \qquad (2.8)$$

Since $\langle \psi|(1/r^2)\psi\rangle \geq 0$, we deduce, for example, monotonicity of the 'Regge trajectories': all energies are increasing functions of the angular momentum ℓ, for a given number of nodes.

Some other consequences are:

Let $\phi_1,\ldots,\phi_N$ be N linear independent, orthonormal, trial functions. Diagonalization of the matrix $\langle\phi_i|H\phi_j\rangle$ yields eigenvalues $\varepsilon_1,\ldots,\varepsilon_N$, which are above the true eigenvalues E_i: $E_i \leq \varepsilon_i$ for $i = 1,\ldots,N$. It follows that

$$\sum_{i=1}^{N} E_i \leq \sum_{i=1}^{N} \langle\phi_i|H\phi_i\rangle .\qquad(2.9)$$

For the sum of the first N eigenvalues, (2.7) also implies

$$\sum_{i=1}^{N} E_i = \inf_{\mathscr{H}_n} \sum_{i=1}^{N} \langle\chi_i|H\chi_i\rangle\qquad(2.10)$$

if $\mathscr{H}_n$ is spanned by $\chi_1,\ldots,\chi_N$ with $\langle\chi_i,\chi_j\rangle = \delta_{ij}$.

The infimum of a family of linear functions is a concave function. This means from (2.6) that E_1 will be a concave function of all parameters entering linearly in H. The same holds for $\sum_{i=1}^{N} E_i$ according to the previous remark.

Clearly, the number of bound states below a certain energy decreases — or at most stays constant — if a positive interaction is added. As a simple consequence we mention that any confining potential V has an infinite number of bound states: one just takes an infinite square well potential V_w such that $V_w \geq V$, and compares the two appropriate Schrödinger operators.

An application of the variational principle tells us that any one-dimensional Schrödinger problem with a potential such that $\int dx V(x)$ is negative has at least one bound state. A similar argument applies to the two-dimensional problem. We shall give the detailed scaling arguments in Section 2.7, when we discuss the possibilities for locating bound states. Such arguments cannot be used for the half-line problem (2.1), since the wave function $u(r)$ has to obey the boundary condition $u(0) = 0$ at the origin. One can actually go from a Schrödinger equation defined on the half-line $r \in [0,\infty)$ to a problem defined on $\mathbf{R}$, by taking $\bar{V}(x) = V(r)$ for $r = x \geq 0$ and $\bar{V}(x) = V(r)$ for $x = -r < 0$, to obtain a symmetric potential $\bar{V}(x) = \bar{V}(-x)$. For each second eigenvalue for $p^2 + V(x)$ the wave function vanishes at the origin $x = 0$, and the eigenvalues for the half-line problem are genuine too.

For the variation of the energy levels as a function of any parameter entering a Hamiltonian $H(\lambda)$ the so-called Feynman–Hellmann theorem is easily obtained:

$$H(\lambda)\psi_\lambda = E\psi_\lambda \Rightarrow \frac{\partial E}{\partial\lambda} = \langle\psi_\lambda|\frac{\partial H(\lambda)}{\partial\lambda}|\psi_\lambda\rangle .\qquad(2.11)$$

Differentiating $E = \langle\psi_\lambda|H(\lambda)\psi_\lambda\rangle$ with respect to λ gives the sum of three expressions; but $\langle \delta\psi_\lambda | H(\lambda)\psi_\lambda \rangle$ and $\langle \psi_\lambda | H(\lambda)\delta\psi_\lambda \rangle$ both vanish because ψ_λ is assumed to be normalized.

A well-known result, the virial theorem, relates expectation values of kinetic energy $T = -d^2/dr^2$ and the 'virial' $r\,dV(r)/dr$:

$$(T + V(r))\psi = E\psi \Rightarrow 2\langle \psi|T\psi \rangle = \left\langle \psi \middle| r\frac{dV}{dr}\psi \right\rangle . \qquad (2.12)$$

Then

$$\delta = \frac{i}{2}\left(r\frac{d}{dr} + \frac{d}{dr}r \right)$$

counts the scaling dimension: $-i[\delta, T] = 2T$ and $-i[\delta, V] = rV'(r)$. The commutator $[\delta, H]$ taken between an eigenstate of H yields Eq. (2.12). In fact, Eq. (2.12) is easily seen to hold for an arbitrary angular momentum.

Power law potentials $V = \lambda r^\alpha$ therefore yield $2\langle T \rangle = \alpha\langle V \rangle$, where $\langle \cdot \rangle \equiv \langle \psi| \cdot \psi \rangle$. In this case we can combine (2.11)

$$\lambda\frac{\partial E}{\partial\lambda} = \langle V \rangle$$

and (2.12), and obtain

$$E = \left(\frac{\alpha}{2} + 1 \right) \lambda\frac{dE}{d\lambda} ,$$

which has the solution $E(\lambda) = \lambda^{2/(\alpha+2)}E(1)$ as long as $\alpha > -2$. A potential such as $-\lambda/r^2$ with $0 < \lambda < 1/4$ is 'scale' invariant but has no bound states. A logarithmic potential $V(r) = \lambda\ln(r/r_0)$ yields, according to (2.12), $2 \cdot \langle \psi|T\psi \rangle = \lambda$.

In the following we shall also consider systems with different quark masses. Therefore, changes of energies with the mass are of interest. If

$$H(m) = -\frac{1}{2m}\frac{d^2}{dr^2} + V(r) ,$$

Eq. (2.11) yields

$$m\frac{dE}{dm} = -\langle T \rangle .$$

For power law potentials

$$E = -\frac{\alpha + 2}{\alpha}\,m\frac{dE}{dm} ,$$

which, integrated, gives $E(m) = m^{-\alpha/(\alpha+2)}E(1)$.

In the special case of $V = \lambda\ln r = \lambda\lim(r^\epsilon - 1)/\epsilon$ we get $m(dE/dm) = -\lambda$, so that dE/dm is independent of the quantum numbers of the state considered. We may quote the scaling behaviour for physical quantities for power law potentials as a function of m, $|\lambda|$ and $\hbar$. We combine our previous assertions and note that m and $\hbar$ enter into (2.2) as a quotient, $\hbar^2/m$, and conclude that

$$E(\lambda, m, \hbar) = |\lambda|^{\frac{2}{\alpha+2}} \left(\frac{\hbar^2}{m}\right)^{\frac{\alpha}{\alpha+2}} E(1, 1, 1) . \qquad (2.13)$$

Familiar cases are the Coulomb potential, where energy levels scale proportionally to $m\lambda^2$, and the harmonic oscillator for which $E \sim (\lambda/m)^{1/2}$. For singular potentials, $\alpha < 0$, Eq. (2.13) implies that level spacings increase with increasing mass, while for regular potentials, $\alpha > 0$, level spacings decrease as a function of m. For the logarithmic potential, scaling shifts the potential only by a constant and energy differences are therefore independent of the mass. If we compare (2.13) and (2.2) we realize that a length, L, has been scaled thus:

$$L \sim \left(\frac{\hbar^2}{m|\lambda|}\right)^{\frac{1}{2+\alpha}} . \qquad (2.14)$$

The decay probability for leptonic processes is proportional to the square of the wave function $|\psi(0)|^2$ at the origin. The scaling behaviour is obtained by noting that $|\psi(0)|^2$ has the dimension of an inverse volume. The total integral of $|\psi(\vec{x})|^2$ over $\mathbf{R}^3$ gives a number. Therefore,

$$|\psi(0)|^2 \sim \left(\frac{m\,|\lambda|}{\hbar^2}\right)^{\frac{3}{2+\alpha}} . \qquad (2.15)$$

A number of useful sum rules and identities are usually obtained by multiplying the Schrödinger equation on both sides by a cleverly chosen function and integrating by parts etc. We have learnt a systematic (and complete) procedure from Bessis [68]: we write (2.2) in the simplified form $u'' = Wu$ and define the density $\rho(r) = u^2(r)$. Differentiation gives $\rho' = 2uu'$, $\rho'' = 2u^2 W + 2u'^2$. The third derivative can be expressed in terms of ρ and ρ' : $\rho''' - 4W\rho' = 0$. We multiply the last equation by a three times differentiable function $F(r)$ and integrate by parts. To simplify, we assume that no boundary terms arise and obtain the most general sum rule

$$\int_0^\infty dr\, \rho(r)\{-F'''(r) + 4F'W + 2FW'\} = 0 . \qquad (2.16)$$

Various special cases of (2.16) will be used later (one special case, the virial theorem, has already been mentioned).

There also exist sum rules, where two different states are involved: we may consider, for example, a wave function which is a superposition of two angular momentum wave functions

$$\phi = \psi_\ell e^{-iE_\ell t} + \psi_{\ell+1} e^{-iE_{\ell+1}t} \qquad (2.17)$$

and insert it into the expression

$$m\frac{d^2}{dt^2}\langle\phi|\vec{r}|\phi\rangle = -\langle\phi|\vec{\nabla}V|\phi\rangle\,, \tag{2.18}$$

which is a special case of Ehrenfest's theorem. We obtain thereby a sum rule of the type

$$\frac{1}{2}(E_{\ell+1} - E_\ell)^2\langle\psi_\ell|r\psi_{\ell+1}\rangle = \left\langle\psi_\ell\left|\frac{dV}{dr}\psi_{\ell+1}\right.\right\rangle\,, \tag{2.19}$$

which will later be used to relate dipole moments (remember our convention $m = 1/2$). In fact these 'mixed' sum rules can be found in a systematic way. If u and v satisfy the equations

$$u'' = Uu\,,$$
$$v'' = Vv\,,$$

one can, by successive differentiations of $uv = \rho$ and using the Wronskian of the two equations, obtain

$$\rho''' = [(U+V)\rho]' + (U+V)\rho' + (U-V)\int_0^r (V-U)\rho(r')dr'\,.$$

Then multiplying by an arbitrary function F and integrating by parts, one gets

$$\int_0^\infty dr\rho[F''' + 2F'(U+V) + F(U+V)' + (V-U)\int_r^\infty (V-U)F(r')dr'] = 0, \tag{2.20}$$

neglecting possible integrated terms.

With the choice $F = 1$, and taking

$$U = W - E_\ell + \frac{\ell(\ell+1)}{r^2}$$

and

$$V = W - E_{\ell+1} + \frac{(\ell+1)(\ell+2)}{r^2}\,,$$

we get back the sum rule (2.19).

2.2 Order of energy levels

As we have already remarked, potentials may be used to describe quarkonium systems. A question raised by Bèg was what leads to the lowest excitation: is it the radial or the orbital? Clearly, the results we obtain can be applied to any domain of physics described by the Schrödinger equation.

The radial Schrödinger equation can be written as

$$\left(-\frac{d^2}{dr^2} + \frac{\ell(\ell+1)}{r^2} + V(r) - E(n,\ell)\right) u_{n,\ell}(r) = 0 . \tag{2.21}$$

For simplicity we shall only deal with 'smooth' potentials, which are less singular than $-1/4r^2$ at the origin so that a ground state exists. The energy eigenvalues $E(n,\ell)$ depend on two quantum numbers, n the number of nodes of the reduced radial wave function $u_{n,\ell}(r)$, and ℓ the orbital angular momentum.

For a completely general potential, which allows for bound states, two properties are well known. First, the energy increases with the number of nodes $E(n+1,\ell) > E(n,\ell)$. This follows from standard Sturm–Liouville theory of second-order differential equations. Second, the energy is an increasing function of the angular momentum $E(n,\ell+1) > E(n,\ell)$. This comes from the fact that the centrifugal potential is repulsive and the coefficient in front of the centrifugal term increases with ℓ. The set of energy levels with the same n and different ℓs form what has been called since 1959 a 'Regge trajectory'. It was pointed out by Regge that ℓ in Eq. (2.21) need not be integer or even real. For real $\ell > -1/2$ we still have $E(n,\ell+\delta) > E(n,\ell)$ for $\delta > 0$.

However, we are interested in a more subtle question. We want to compare energy levels for pairs of (n,ℓ) and (n',ℓ'). In order to get an initial insight we quote the well-known special cases with exceptional degeneracy:

The Coulomb potential

If $V(r) = -\alpha/r$, the energy eigenvalues depend only on the principal quantum number $N = n + \ell + 1$, and we get the energy level structure shown in Figure 2.1. Here we have $E(n+1,\ell) = E(n,\ell+1)$.

The harmonic oscillator potential

If $V(r) = \omega^2 r^2/2$, the energy eigenvalues depend only on the combination $n + \ell/2$. One has to go up by two units in angular momentum to reach the next degenerate level: $E(n+1,\ell) = E(n,\ell+2)$. The energy level structure is shown in Figure 2.2. Regge trajectories are indicated by dotted lines. Straight lines represent the harmonic oscillator, and concave ones the Coulomb potential trajectories.

Next we shall state and prove one of our main results concerning the energy level ordering, also mentioned in (1.14). Typically, we impose conditions on the potential. We compare the energy level scheme with that of the examples given previously.

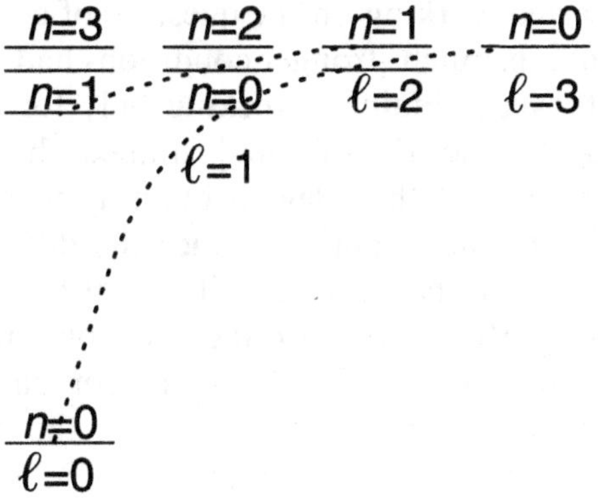

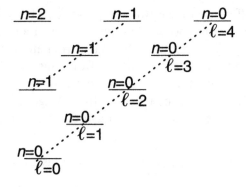

Fig. 2.1.   Energy level structure of the Coulomb Schrödinger operator.

Fig. 2.2.   Energy level structure of the harmonic oscillator Schrödinger operator.

Theorem:

Assume that the Laplacian of the potential (away from $r = 0$) has a definite sign,

$$\Delta V(r) = \frac{1}{r^2} \frac{d}{dr} r^2 \frac{dV(r)}{dr} \gtrless 0 \quad \forall r > 0 .$$

The energy levels will then be ordered

$$E(n + 1, \ell) \gtrless E(n, \ell + 1) . \tag{2.22}$$

Remarks:

Naturally, in the Coulomb case, $\Delta V = 0$ outside the origin, which means that the result is in its optimal form.

A slight refinement is that $\Delta V \leq 0$, or $dV/dr \leq 0$, for $r \neq 0$ is already enough to guarantee $E(n + 1, \ell) \leq E(n, \ell + 1)$. A number of variants of Theorem (2.22) will be mentioned later.

Historically we started working on the question of energy level ordering in 1977. At the beginning, much stronger conditions had to be imposed and much weaker results were obtained, working only for $n = 0$ and $n = 1$. Then, Feldman, Fulton and Devoto [69] studied the WKB limit with $n \gg \ell \gg 1$. They found that the relevant quantity was $(d/dr)r^2(dV/dr)$, which is proportional to the Laplacian. Since the difference between the WKB approximation and the exact result is hard to control, we next looked at the case of perturbations around the Coulomb (and oscillator) potential: $V(r) = -(\alpha/r) + \lambda v(r)$. In order to compare $E(n+1,\ell)$ and $E(n,\ell+1)$ in the limit $\lambda \to 0$, we need to find the sign of the quantity

$$\delta = \int_0^\infty dr\, v(r)[u_{n+1,\ell}^2(r) - u_{n,\ell+1}^2(r)]\,, \tag{2.23}$$

which gives the difference of the two energy shifts. The us in Eq. (2.23) are the unperturbed Coulomb wave functions — i.e., Laguerre polynomials times exponentials. The integrand oscillates tremendously if n is large. However, $u_{n+1,\ell}$ and $u_{n,\ell+1}$ are obtained from one another by applying a raising or a lowering operator and the existence of these operators is connected to the O_4 symmetry and the Runge–Lenz vector in the Coulomb problem. They are simple first-order differential operators of the form

$$u_{n,\ell+1} = \text{const}\, A_+\, u_{n+1,\ell}\,, \quad u_{n+1,\ell} = \text{const}\, A_-\, u_{n,\ell+1}\,,$$

$$A_\pm = \pm\frac{d}{dr} - \frac{\ell+1}{r} + \frac{1}{2(\ell+1)}\,. \tag{2.24}$$

All that needed to be done was to replace $u_{n+1,\ell}$ according to (2.24) and integrate twice by parts. In this way, δ of Eq. (2.23) appears as an integral over the Laplacian of v times a positive quantity. Therefore, the sign of the Laplacian of the potential determines the way levels are split in first-order perturbation theory [70].

The problem was then to get a non-perturbative result, which was solved in Ref. [71]. The idea was to generalize the notion of the raising operator. If $u_{n,\ell}$ is a solution of Eq. (2.21), and defining

$$\tilde{u} = \left(\frac{d}{dr} - \frac{u'_{0,\ell}}{u_{0,\ell}}\right)u_{n,\ell} = \frac{u_{0,\ell}u'_{n,\ell} - u'_{0,\ell}u_{n,\ell}}{u_{0,\ell}}\,,$$

where $u_{0,\ell}$ denotes the ground-state wave function, with $n = 0$ and angular momentum ℓ, then $\tilde{u}$ is a solution of

$$\left[-\frac{d^2}{dr^2} + \frac{\ell(\ell+1)}{r^2} + 2\frac{d}{dr}\left(-\frac{u'_{0,\ell}}{u_{0,\ell}}\right) + V - E\right]\tilde{u} = 0. \tag{2.25}$$

This transformation, which we reinvented, has actually been known for a long time. It entered the work of Marchenko and Crum in 1955, but in fact goes back to Darboux, who studied it in 1882. Its relation to supersymmetric quantum mechanics is explained in Appendix A.

If V is regular at the origin, u behaves like $r^{\ell+1}$, and we find from the definition that $\tilde{u}$ behaves like $r^{\ell+2}$. In addition, let us show that $\tilde{u}$ has one node less than u: between two successive zeros of u, $\tilde{u}$ has at least one zero, since u' and hence $\tilde{u}$ have opposite signs at the zeros. We can combine (2.21) for $u_0 = u_{0,\ell}$ and energy E_0, and $u = u_{n,\ell}$ with energy E, to get, after integration,

$$u_0(r_2)\tilde{u}(r_2) - u_0(r_1)\tilde{u}(r_1) = (E - E_0) \int_{r_1}^{r_2} dr\, u_0(r)u_{n,\ell}(r)\,. \tag{2.26}$$

From (2.26) we see that between two successive zeros of $\tilde{u}$ there is a zero of $u_{n,\ell}$, remembering that u_0 has a constant sign. By taking $r_1 = 0$ we find that, as long as $u_{n,\ell}$ is, say, positive, $\tilde{u}$ is positive and has a behaviour like $r^{\ell+2}$ at the origin if u_0 and $u_{n,\ell}$ behave like $cr^{\ell+1}$. Therefore $\tilde{u}$ has the characteristic behaviour corresponding to angular momentum $\ell + 1$. By taking $r_2 \to \infty$ we find that beyond the last zero of $u_{n,\ell}$, $\tilde{u}$ has a constant sign opposite to that of $u_{n,\ell}$. Hence $\tilde{u}$ has one node less than $u_{n,\ell}$. Therefore $\tilde{u}$ and $u_{n,\ell}$ correspond to the same principal quantum number, $N = n + \ell + 1$.

We can consequently interpret $\tilde{u}$ as a wave function with angular momentum $\ell + 1$ and $n - 1$ nodes feeling a potential

$$\tilde{V} = V + 2\left[\frac{d}{dr}\left(-\frac{u'_{0,\ell}}{u_{0,\ell}}\right) - \frac{\ell+1}{r^2}\right]. \tag{2.27}$$

Next comes a crucial lemma which allows us to tell under which conditions $\tilde{V} \geq V$ or $\tilde{V} \leq V$.

Lemma:

$$\text{If} \quad \Delta V(r) \gtrless 0 \quad \forall r > 0\,,$$

$$\frac{d}{dr}\left(-\frac{u'_{0,\ell}}{u_{0,\ell}}\right) - \frac{\ell+1}{r^2} \gtrless 0\,, \tag{2.28}$$

$$\text{i.e,} \quad \frac{d^2}{dr^2}\log\left(\frac{u_{0,\ell}}{r^{\ell+1}}\right) \lessgtr 0\,.$$

In fact, for the second case ($\tilde{V} \leq V$) it is actually enough that $\Delta V \leq 0$

or $dV/dr \leq 0$ for all $r > 0$, in order to conclude that

$$\frac{d}{dr}\left(-\frac{u'_{0,\ell}}{u_{0,\ell}}\right) - \frac{\ell+1}{r^2} \leq 0.$$

For this refinement we direct the reader to the original reference [71].

Note:

Clearly for the Coulomb case we obtain equalities. For the first proof of the lemma we used the Coulomb potential V_c as a comparison potential such that $E - V(R) = E_c - V_c(R)$, and $dV/dr(R) = dV_c/dr(R)$, where E_c is the ground-state energy of the Coulomb problem. Then, if V is, for example, convex in $1/r$, $E - V$ and $E_c - V_c$ do not intersect in the interval $(0, R)$ nor in (R, ∞). This allowed the writing of a Wronskian relation for u and u_c, which gave the result.

The new proof is even simpler. The steps are motivated by the result of Ashbaugh and Benguria [72]. We return to Eq. (2.25) and consider $\tilde{u}$ as a reduced wave function with $n - 1$ nodes and angular momentum $\ell + 1$, which satisfies a Schrödinger equation with a potential $V + \delta V$, where

$$\frac{1}{2}\delta V = -\left(\frac{u'}{u}\right)' - \frac{\ell+1}{r^2} ; \tag{2.29}$$

$u_{0,\ell}$ is denoted by u.

We intend to show that δV is positive (or negative) everywhere if $rV(r)$ is convex (or concave). In Ref. [72] it is remarked that the result of the lemma is equivalent to saying that $v = \ln(u/r^{\ell+1})$ is concave (or convex) when the Laplacian of the potential is positive (or negative). Following Ref. [73] we work directly with $\delta V = -2v''$ and obtain from the Schrödinger equation

$$u^2\frac{\delta V}{2} = (E - V)u^2 + u'^2 - \frac{(\ell+1)^2}{r^2}u^2$$

and get, after two differentiations,

$$\frac{d}{dr}\frac{r^2}{u^2}\frac{d}{dr}u^2\delta V = -\frac{d}{dr}\left(r^2\frac{dV}{dr}\right) + 2(\ell+1)\delta V , \tag{2.30}$$

where we recognize the three-dimensional Laplacian on the r.h.s. of (2.30). Notice that

$$\lim_{r\to 0} u^2\delta V = 0 , \tag{2.31}$$

at least for potentials with $\lim_{r\to 0} r^2V(r) = 0$, because then $u \sim r^{\ell+1}$. In addition, $\lim_{r\to\infty} u^2\delta V = 0$, which follows because potentials with a definite Laplacian are necessarily monotonous beyond a certain value of r. Assume now that $\Delta V \geq 0$. From (2.30) we obtain a linear differential inequality for δV. Assume that somewhere between 0 and ∞ $\delta V(r)$ is negative. This means that $u^2 \delta V$ has at least one minimum $\bar{r}$, where δV is negative. At this minimum, the l.h.s. of (2.30) is positive, while the r.h.s. is negative. This is a contradiction. Therefore, δV is positive everywhere.

Now take, for example, $\Delta V \geq 0$. Then we have $\widetilde{V} > V$. However, since the energy remained the same in going from V to $\widetilde{V}$ we have $E_{n,\ell}(V) = E_{n-1,\ell+1}(\widetilde{V})$. Because the energies are monotonous in the potential we obtain $E_{n-1,\ell+1}(\widetilde{V}) \geq E_{n-1,\ell+1}(V)$, which proves the result. Clearly, the same argument applies to the opposite sign. We therefore obtain a large class of potentials for which the levels which are degenerate in the Coulomb case are split in a very definite way.

Applications:

As mentioned in the introduction we have used this result to analyse both quarkonium and muonic systems. We have also based a new interpretation of the energy level structure of atomic systems as it determines the periodic table on this knowledge of how energy levels split [74]. In quarkonium and muonic systems higher angular momentum states are lower and the upper sign of (2.2) applies, and it can be argued that in the case of atoms just the opposite holds. In muonic atoms, the muon comes so close to the nucleus that the interaction with electrons becomes negligible, while the finite extension of the nucleus becomes relevant. The muon feels the non-negative charge distribution with a potential

$$V(\vec{x}) = -e^2 \int \frac{d^3 y \rho(\vec{y})}{|\vec{x} - \vec{y}|}, \tag{2.32}$$

which is spherically symmetric. Since ΔV is positive, the same energy level order as in quarkonium shows up.

To discuss some of the spectra of atomic systems, we treat the alkaline atoms in the Hartree approximation. These can be treated as the interaction of an outer electron with a spherically symmetric charge distribution. We quote, first, the standard argument, which is mentioned also in Condon and Shortley [75]. It is argued for sodium, for example, that the 3S state is lower than the 3P state, because the electron will be closer to the nucleus for the former state. But this argument is not convincing. It could be applied to the hydrogen atom too, but there we know that both energies become degenerate. We shall put our arguments on a different basis and consider the Schrödinger equation for an external electron interacting with a charge distribution ρ

$$-\Delta \psi(\vec{x}) - \frac{Z e^2}{|\vec{x}|} \psi(\vec{x}) + e^2 \int d^3 y \frac{\rho(\vec{y})}{|\vec{x} - \vec{y}|} \psi(\vec{y}) = E \psi(\vec{x}), \tag{2.33}$$

where $\rho(\vec{y}) = \sum_i |\psi_i(\vec{y})|^2$ should be self-consistently determined from the non-linear Eq. (2.33). But since ρ is positive, the Laplacian of the effective potential is negative. We therefore have $E(n+1, \ell) < E(n, \ell+1)$ and deduce the ordering

$$E(3S) < E(3P) < E(3D),$$

$$\tag{2.34}$$

$$E(4S) < E(4P) < E(4D) < E(4F),$$

which is verified by explicit calculations [76].

Within the Hartree approximation and neglecting the interaction between outer electrons, we may even use the above result to justify the filling of levels in atoms. And (2.34) is indeed fulfilled. In argon the 3P shell is filled, and then the question arises as to whether the next electron will be in a 3D state or in a 4S state. It happens that the 4S state wins. This depends clearly on the fine details of the interaction and is not in contradiction with our inequalities. So it appears that no violation to our deduced ordering occurs.

Together with Baumgartner we applied the method that allowed us to order levels to the continuous spectrum as well. There we found relations between scattering phase shifts [77]. Clearly, we had to be careful to take into account the asymptotic behaviour of the wave functions at infinity. We compared, therefore, phase shifts relative to a Coulomb potential $-Z/r$. Here we quote only the result.

We assume that $V(r) = -Z/r + V_{SR}(r)$, where V_{SR} denotes a short-range potential and also that V_{SR} fulfils

$$\lim_{r \to 0} r^2 V(r) > -\frac{1}{4}, \quad \int_{r_0}^{\infty} dr |V_{SR}(r)| < \infty, \quad V_{SR}(\infty) = 0 \tag{2.35}$$

and $\Delta V \leq 0$ for $r \neq 0$. We denote by $\delta_\ell(E)$ the ℓ-th scattering phase shift relative to the Coulomb potential. Then

$$\delta_{\ell+1}(E) \leq \delta_\ell(E). \tag{2.36}$$

Moreover, if the ground-state energy for angular momentum ℓ is lower than $-(Z/2(\ell+1))^2$ and $Z \leq 0$, or if $Z > 0$ we get

$$\delta_{\ell+1}(E) \leq \delta_\ell(E) + \arctan\frac{Z}{2k(\ell+1)} - \arctan\frac{\sqrt{|E_\ell|}}{k}, \tag{2.37}$$

where $E = k^2$, and E_ℓ is the infimum of the spectrum of (2.1).

If now (2.35) holds, and $\Delta V \geq 0$ for all $r \neq 0$ is fulfilled, $\delta_\ell(E)$ is monotonous in ℓ: $\delta_{\ell+1}(E) \geq \delta_\ell(E)$. For $Z < 0$, E_ℓ denotes the ground-state energy with $\sqrt{|E_\ell|} \leq Z/2(\ell+1)$ and we get

$$\delta_{\ell+1}(E) \geq \delta_\ell(E) + \arctan\frac{Z}{2k(\ell+1)} - \arctan\frac{\sqrt{|E_\ell|}}{k}, E = k^2. \tag{2.38}$$

Generalizations:

It would be surprising if starting from the energy level scheme of an oscillator, no analogous results could be obtained. In order to get these results one has to remember that there exists a transformation which allows change from the Coulomb problem to the harmonic oscillator. Since in the latter case the n-th level of angular momentum ℓ is degenerate with the $(n-1)$-th level of angular momentum $\ell+2$, it is expected that the splitting of these levels will replace (2.2). Such a result was first known to us perturbatively [70].

There exists an even more general result [78]. We found local conditions on the potential such that the n-th level to angular momentum ℓ lies above (or below) the $(n-1)$-th level of angular momentum $\ell+\alpha$ for any given α. One of these conditions says that for a *positive* potential with

$$D_\alpha V(r) > 0, \qquad 1 < \alpha < 2 \Rightarrow E(n,\ell) > E(n-1,\ell+\alpha),$$
$$\tag{2.39}$$

$$D_\alpha V(r) > 0, \quad 2 < \alpha \quad \text{or} \quad \alpha < 1 \Rightarrow E(n,\ell) < E(n-1,\ell+\alpha),$$

where the second-order differential operator D_α is defined by

$$D_\alpha = \frac{d^2}{dr^2} + (5 - 3\alpha)\frac{1}{r}\frac{d}{dr} + 2(1-\alpha)(2-\alpha)\frac{1}{r^2}. \tag{2.40}$$

Let us mention especially the case $\alpha = 2$. Convexity (respectively concavity) of the potential in r^2 implies relations between energy levels:

$$\left(\frac{d}{dr^2}\right)^2 V(r \gtrless 0 \Rightarrow E(n,\ell) \gtrless E(n-1,\ell+2). \tag{2.41}$$

Note that equality for the energy levels on the r.h.s. of (2.41) for all n and ℓ is obtained for the harmonic oscillator. Equation (2.41) is a generalization of our previous result concerning perturbations around the harmonic oscillator. The well-known transformation from the Coulomb problem to the harmonic oscillator lies at the origin of these results.

The generalization which we mentioned is formulated in the following theorem.

Theorem:

Assume that $V(r)$ is positive, and $D_\alpha V(r)$ is positive for $1 < \alpha < 2$ (negative for $\alpha > 2$ or $\alpha < 1$), where D_α has been given in (2.40). Then

$$E(n,\ell) \gtrless E(n-1,\ell+\alpha). \tag{2.42}$$

Similarly, if $V(r)$ is *negative* and $D_\alpha V(r)$ is negative for $1 < \alpha < 2$ (positive for $\alpha < 1$) (note that in the case $D_\alpha V(r) > 0$, $\alpha > 2$, $V(r) < 0$ is empty!), then

$$E(n,\ell) \gtrless E(n-1, \ell+\alpha). \tag{2.43}$$

Proof:

The main idea is to transform the Schrödinger equation in such a way as to be able to apply the techniques explained previously: starting from (2.2) we make the change of variables from r to z and change of wave function from $u_{n,\ell}$ to $w_{n,\ell}$:

$$z = r^\alpha, \qquad w_{n,\ell}(z) = r^{\frac{\alpha-1}{2}} u_{n,\ell}(r), \qquad W(z) = V(r), \qquad (2.44)$$

which is a special case of a more general transformation [79] and gives

$$\frac{d^2}{dz^2} w_{n,\ell}(z) = \left\{ U(z; E(n,\ell)) + \frac{\lambda(\lambda+1)}{z^2} \right\} w_{n,\ell}(z),$$

$$U(z; E) = \frac{W(z) - E}{\alpha^2 z^{2-2/\alpha}}, \qquad (2.45)$$

where $\lambda = (2\ell - \alpha + 1)/2\alpha$ denotes the new angular momentum. Equation (2.45) can be considered as a Schrödinger equation with angular momentum λ, potential $U(z; E)$ and energy *zero*. From the assumed positivity of $E(n,\ell)$ and $D_\alpha V(r)$ for $1 < \alpha < 2$ we deduce by straightforward computation that $\Delta_z U(z; E) > 0$; therefore there exists a new potential $\tilde{U}$, which has a zero-energy state of angular momentum $\lambda + 1$, and a wave function which has one node less than $w_{n,\ell}$. Returning to the initial Schrödinger equation (2.2) one gets a state with angular momentum $\ell + \alpha$ and a potential which is above the old one. The new potential now depends on n and $E(n,\ell)$, but that does not matter. We deduce, again using the min-max principle, that $E(n,\ell) > E(n-1, \ell + \alpha)$. The reversed sign for $\alpha > 2$ or $\alpha < 1$ follows from the assumption that $D_\alpha V(r) < 0$ for $\alpha > 2$ or $\alpha < 1$.

The proof of the second part of this theorem is similar. ∎

Examples:

A source of examples is given by pure power potentials $\epsilon(v) r^v$, where ϵ is the sign function. Then $D_\alpha V(r)$ reduces to $\epsilon(v)[v - 2(\alpha-1)][v + 2 - \alpha] r^{v-2}$. In this way, we get

$$\begin{aligned}
V &= r^4, & E(n-1, \ell+2) &< E(n,\ell) < E(n-1, \ell+3), \\
V &= r, & E(n-1, \ell+3/2) &< E(n,\ell) < E(n-1, \ell+2), \\
V &= -r^{-1/2}, & E(n-1, \ell+1) &< E(n,\ell) < E(n-1, \ell+3/2) \\
V &= -r^{-3/2}, & E(n-1, \ell+1/2) &< E(n,\ell) < E(n-1, \ell+1).
\end{aligned} \qquad (2.46)$$

Remark:

We observe that there are potentials which solve the differential equation $D_\alpha V(r) = 0$ and which are transformed into Coulomb potentials.

Theorem:

The potentials

$$V_{\alpha,N,Z}(r) = \alpha^2 \frac{Z^2}{4N^2} r^{2\alpha-2} - \alpha^2 Z r^{\alpha-2}, \quad \alpha > 0, \ Z > 0, \ N \geq \frac{1}{2} \quad (2.47)$$

have zero-energy eigenvalues for angular momentum quantum numbers ℓ which satisfy

$$\lambda = \frac{2\ell - \alpha + 1}{2\alpha} = N - n - 1, \quad n = 0, 1, 2, \dots, \quad (2.48)$$

where n denotes the number of nodes of the corresponding eigenfunction.

Proof:

With a change of variable from r to $\rho = r^\alpha$, the radial Schrödinger equation with potential $V_{\alpha,N,Z}$, angular momentum ℓ and energy zero is transformed into the Schrödinger equation for the Coulomb potential $-Z/\rho$, angular momentum λ and energy $-Z^2/4N^2$. It remains to remark that the Coulomb quantum number N need not be integer: the algebraic treatment of the Coulomb problem [67, 78] works for general real angular momentum λ. ∎

Remark:

One may also do a direct algebraic treatment for the potentials described in theorem (2.47). Let us define

$$A^\pm_{\alpha,Z,\ell} = \pm \frac{1}{\alpha} r^{1-\alpha} \frac{d}{dr} - \frac{\mu_\pm(\alpha)}{\alpha r} + \frac{\alpha Z}{2\ell + 1 + \alpha}, \quad \mu_+ = \ell + 1, \ \mu_- = \ell + \alpha, \quad (2.49)$$

$$H_{\alpha,N,Z,\ell} = -\frac{d^2}{dr^2} + \frac{\ell(\ell+1)}{r^2} + V_{\alpha,N,Z}(r), \quad (2.50)$$

with Dirichlet boundary conditions at $r = 0$. One calculates easily, on the one hand, the product of A^- and A^+:

$$A^-_{\alpha,Z,\ell} A^+_{\alpha,Z,\ell} = \frac{1}{\alpha^2} r^{2-2\alpha} H_{\alpha,N(\ell),Z,\ell} \quad \text{with} \quad N(\ell) = \frac{2\ell + 1 - \alpha}{2\alpha} + 1, \quad (2.51)$$

and, on the other hand, that the commutator is

$$\left[A^+_{\alpha,Z,\ell}, A^-_{\alpha,Z,\ell} \right] = \frac{1}{\alpha^2} r^{2-2\alpha} (H_{\alpha,N,Z,\ell+\alpha} - H_{\alpha,N,Z,\ell}). \quad (2.52)$$

This shows that

$$r^{2-2\alpha} H_{\alpha,N,Z,\ell-\alpha} A^-_{\alpha,Z,\ell-\alpha} = A^-_{\alpha,Z,\ell-\alpha} r^{2-2\alpha} H_{\alpha,N,Z,\ell}. \quad (2.53)$$

The ground-state wave function $u_{0,\ell}$ for $H_{\alpha,N,Z,\ell}$ is the solution of the differential equation $A^+_{\alpha,Z,\ell}u_{0,\ell} = 0$ and is given explicitly by

$$u_{0,\ell}(r) = r^{\ell+1}e^{-Zr^\alpha/2N} \ . \tag{2.54}$$

Furthermore, proceeding inductively we see that

$$u_{n,\ell-n\alpha} = A^-_{\alpha,Z,\ell-n\alpha}u_{n-1,\ell-(n-1)\alpha} \tag{2.55}$$

is a solution of

$$H_{\alpha,N,Z,\ell-n\alpha}u_{n,\ell-n\alpha} = 0 \ .$$

The case $\alpha = 2$, which is a comparison to the level ordering for the three-dimensional harmonic oscillator, is special.

Corollary:

$$\begin{aligned} \text{If} \quad D_2V(r) \geq 0 \quad &\text{then} \quad E(n,\ell) \geq E(n-1,\ell+2) \ , \\ \text{if} \quad D_2V(r) \leq 0 \quad &\text{then} \quad E(n,\ell) \leq E(n-1,\ell+2) \ . \end{aligned} \tag{2.56}$$

Proof:
Since D_2 contains no zero-order term, the conditions $D_2V(r)\gtrless 0$ are equivalent to $\Delta U \gtrless 0$ when we make the same transformations as in the proof of the foregoing theorem. Therefore, both of the inequalities for the levels of U yield the inequalities for the levels of V. ∎

Remark:
One can define a 'running elastic force constant' $K(r)$ by $V'(r) = K(r)r$. The conditions $D_2V(r) > (<)0$ are then equivalent to the monotonicity of $K : K'(r) > (<)0$. This is analogous to the case $\alpha = 1$, in which we can consider a 'running Coulomb constant' $Z(r)$ defined by $V'(r) = Z(r)/r^2$. Then, $\Delta V(r) > (<)0$ means $Z'(r) > (<)0$.

Remark:
In Theorem (2.42) one may avoid the assumption that V is positive and reformulate it in terms of $U(z; E)$; but then the condition is energy-dependent. Eliminating the energy dependence in such a condition leads to the next results.

Theorem:
If the potential $V(r)$ fulfils

$$\left(r\frac{d^2}{dr^2} + (3 - 2\alpha)\left(\frac{d}{dr}\right)\right)V(r) \leq 0 \ \text{for} \ \alpha \geq 2 \ , \tag{2.57}$$

we deduce that $E(n,\ell) \leq E(n-1,\ell+\alpha)$.

Proof:

We start from the transformed Schrödinger equation (2.45) in the variable $z = r^\alpha$ and use a refined version of our previously derived theorem (2.42)–(2.43) [71]. There exists a solution to (2.45), with λ increased by one unit and energy zero and a potential lower than U if

$$I(z) = \Delta_z U(z; E) < 0 \quad \text{wherever} \quad J(z) = \frac{d}{dz} U(z; E) > 0 . \tag{2.58}$$

$J(z)$ positive means explicitly that

$$z \frac{d}{dz} W(z) > \left(2 - \frac{2}{\alpha}\right) (W(z) - E) . \tag{2.59}$$

Now it is simple to realize that (2.59) together with negativity of

$$K(z) = z \frac{dW}{dz} \left(\frac{2}{\alpha} - 1\right) + z^2 \frac{d^2}{dz^2} W < 0 , \tag{2.60}$$

implies that $I(z)$ is negative for $\alpha > 2$. Note that

$$\alpha z I(z) = \alpha z^{2/\alpha - 3} K(z) + (2 - \alpha) J(z) . \tag{2.61}$$

Condition (2.60) rewritten in the variable r gives (2.57). ∎

Remark:

For pure power potentials we get nothing new compared to Theorem (2.43), but condition (2.57) is — as it should be — invariant under a change of origin of energies.

Since we did not find a way to weaken the condition $\Delta U(z; E) > 0$, removing the energy dependence in that case can be done only with the help of an additional assumption.

Theorem:

Let $V(r)$ be monotonously increasing $dV/dr > 0$, and assume

$$\left(r \frac{d^2}{dr^2} + (3 - 2\alpha) \frac{d}{dr}\right) V(r) \geq 0 \quad \text{with} \quad 1 < \alpha < 2 , \tag{2.62}$$

then

$$E(n, \ell) > E(n - 1, \ell + \alpha) . \tag{2.63}$$

Proof:

This time we assume that $K(z)$ is positive, which, rewritten in the variable r, means (2.62). In order again to use (2.62) and to deduce that $I(z)$ is positive for $\alpha > 2$, we would like to show the positivity of $J(z)$. This can be obtained by noting that $E(n,\ell) > W(0)$ from the monotonicity of the potential, and $W(z) - W(0)$ can be bounded, since

$$W(z) - W(0) = \frac{\alpha}{2(\alpha - 1)} \left\{ z W' - \int_0^z dy \frac{K(y)}{y} \right\} \qquad (2.64)$$

and K is positive. This shows that

$$W(z) - E(n,\ell) < \frac{\alpha}{2(\alpha - 1)} z W'(z) \qquad (2.65)$$

and implies positivity of $J(z)$ and $I(z)$.

Finally we formulate the result that follows.   ∎

Theorem:

Assume that $V(r)$ fulfils

$$r \frac{d^2 V}{dr^2} + (3 - \alpha) \frac{dV}{dr} \leq 0 \quad \text{for } 1 < \alpha < 2, \qquad (2.66)$$

then $E(n,\ell) < E(n - 1, \ell + \alpha)$.

Proof:

This time we observe that

$$E(n,\ell) - W(z) < \int_z^\infty dy\, W'(y) = \frac{\alpha}{2 - \alpha} \left\{ z W' + \int_z^\infty dy \left[y W'' + \frac{2}{\alpha} W' \right] \right\}. \qquad (2.67)$$

Assume now that

$$\left(\alpha z \frac{d^2}{dz^2} + 2 \frac{d}{dz} \right) W(z) < 0, \qquad (2.68)$$

which is equivalent to (2.66); this shows that

$$E(n,\ell) - W(z) < \frac{\alpha}{2 - \alpha} z W'(z) . \qquad (2.69)$$

Now splitting $I(z)$ from (2.58) in an obvious way so as to eliminate the term proportional to $E(n,\ell) - W(z)$, we deduce that $\Delta_z U(z; E(n,\ell))$ is negative by again using (2.68).   ∎

Conjectures:

Besides the expected result for the comparison with the order of energy levels of the harmonic oscillator we have obtained a series of new and unexpected results, illustrated by the examples (2.47). However, when one tests these inequalities numerically one sees that they are not as constraining as one might expect. For instance, for the potential $V = \ln r$, which corresponds to $\nu \to 0$, we do not get anything new, but just

$E(n-1, \ell+1) < E(n, \ell) < E(n-1, \ell+2)$. We may ask whether the factors $(3 - 2\alpha)$ in conditions (2.57) and (2.62), as well as the factor $(3 - \alpha)$ in condition (2.66), are optimal or could be replaced by different factors. Based on the harmonic oscillator approximation for large angular momenta and on numerical checks we are led to formulate improved versions of Theorems (2.57) and (2.62)–(2.66) as conjectures.

Conjecture:
Assume that $V(r)$ fulfils

$$\left(r \frac{d^2}{dr^2} - (\alpha^2 - 3) \frac{d}{dr} \right) V(r) < 0 \quad \text{for } \alpha > 2, \ \alpha < 1. \tag{2.70}$$

We conjecture that $E(n, \ell) < E(n-1, \ell+\alpha)$. If, on the other hand,

$$\left(r \frac{d^2}{dr^2} - (\alpha^2 - 3) \frac{d}{dr} \right) V(r) < 0 \quad \text{for } 1 < \alpha < 2, \tag{2.71}$$

we conjecture the reverse ordering $E(n, \ell) > E(n-1, \ell+\alpha)$.
We have a number of arguments supporting this conjecture.

First argument:
For smooth potentials and large ℓ, we expect that the effective potential

$$V_\ell(r) = \frac{\ell(\ell+1)}{r^2} + V(r) \tag{2.72}$$

will, near its minimum, look more and more like an harmonic oscillator. With r_ℓ being the place of the minimum of $V_\ell(r)$:

$$2 \frac{\ell(\ell+1)}{r_\ell^3} = V'(r_\ell), \tag{2.73}$$

we expect that for ℓ going to infinity with n staying finite the energy levels will be determined by the harmonic oscillator frequencies around r_ℓ:

$$E(n, \ell) \simeq V_\ell(r_\ell) + (2n+1) \sqrt{\frac{1}{2} V_\ell''(r_\ell)}, \tag{2.74}$$

where the curvature is given by

$$V_\ell''(r_\ell) = \frac{6\ell(\ell+1)}{r_\ell^4} + V''(r_\ell) = 3 \frac{V'(r_\ell)}{r_\ell} + V''(r_\ell). \tag{2.75}$$

Since one expects for smooth potentials that for $\ell \to \infty$ and $r_\ell \to \infty$ [60]

$$V' \simeq 0 \left(\frac{V}{r_\ell} \right), \quad V'' = 0 \left(\frac{V}{r_\ell^2} \right), \quad V_\ell \simeq= 0 \left(\frac{\ell^2}{r_\ell^2} \right), \quad \sqrt{V_\ell''} \simeq 0 \left(\frac{\ell}{r_\ell^2} \right), \tag{2.76}$$

the leading term in (2.74) will be given by V_ℓ. Substituting $\ell + \Delta\ell$ for ℓ into (2.74) and using

$$V_\ell(r_{\ell+\Delta\ell}) - V_\ell(r_\ell) \simeq (r_{\ell+\Delta\ell} - r_\ell)\frac{d}{dr}V_\ell(r_\ell) = 0, \qquad (2.77)$$

which holds at the minimum, gives for the energy difference

$$E(n, \ell + \Delta\ell) - E(n, \ell) \simeq V_{\ell+\Delta\ell}(r_\ell) - V_\ell(r_\ell) \simeq \frac{(2\ell+1)\Delta\ell}{r_\ell^2}. \qquad (2.78)$$

One may now ask for which value of $\Delta\ell$, does $E(n, \ell)$ equal $E(n-1, \ell+\Delta\ell)$. This will be the case if the r.h.s. of (2.78), which can be related to $V'(r_\ell)$ by using (2.73), equals the contribution from changing n to $n-1$ in (2.74). With the help of (2.75) we get

$$2\Delta\ell\sqrt{\frac{V'(r_\ell)}{2r_\ell}} \simeq \frac{(2\ell+1)\Delta\ell}{r_\ell^2} = 2\sqrt{\frac{3V'(r_\ell) + r_\ell V''(r_\ell)}{r_\ell}}. \qquad (2.79)$$

Therefore, we deduce the equality of the energy levels in the limit we are considering if

$$r_\ell V''(r_\ell) = \left[(\Delta\ell)^2 - 3\right]V'(r_\ell). \qquad (2.80)$$

Furthermore, if the l.h.s. of (2.80) exceeds the r.h.s. one expects $E(n, \ell) > E(n-1, \ell + \Delta\ell)$ and vice versa for large ℓ. Since for smooth potentials going to infinity for $r \to \infty$, we expect the WKB approximation to be valid for ℓ finite, $n \to \infty$, we expect at least for confining potentials that $E(n, \ell) \simeq E(n-1, \ell+2)$ [69] and only one of the inequalities will survive for $\Delta\ell > 2$ or $\Delta\ell < 2$ for $n \to \infty$, ℓ finite. Thus we expect (2.70) and (2.71) to hold.

Second argument:
For pure power potentials we found an asymptotic expansion of $E(n, \ell)$ in decreasing powers of $\ell + 1/2$. First, we again approximate

$$V_\ell(r) = \frac{\ell(\ell+1)}{r^2} + \epsilon(v) \cdot r^v, \quad \epsilon(v) = \begin{cases} +1, & v > 0 \\ -1, & v < 0 \end{cases} \qquad (2.81)$$

by an harmonic oscillator and calculate anharmonic corrections. It turns out that (for pure power potentials) the first-order perturbation due to the term $V_\ell^{IV}(r_\ell)(r-r_\ell)^4/4!$ is of the same order as the second-order perturbation due to $V_\ell^{III}(r_\ell)(r-r_\ell)^3/3!$. It is tedious but simple to calculate matrix elements of x^3 and x^4 using the $(a + a^\dagger)$ representation of x. (A similar expansion can be found in Ref. [80].) The final answer is

Table 5. Energy levels for the potential $V(r) = r^4$ for the angular momentum ℓ and $\ell + \sqrt{6}$.

ℓ	0	1	2	3	n 4	5	6	7
0	3.8	11.6	21.2	32.1	44.0	56.7	70.3	84.5
$\sqrt{6}$	12.6	22.8	34.0	46.2	59.2	72.9	87.3	102.3
1	7.1	16.0	26.4	37.8	50.1	63.3	77.2	91.7
$1+\sqrt{6}$	16.9	27.7	39.5	52.2	65.6	79.7	94.4	109.7
2	10.8	20.6	31.6	43.6	56.4	69.9	84.2	99.0
$2+\sqrt{6}$	21.4	32.8	45.2	58.3	72.1	86.6	101.6	117.2
3	14.9	25.5	37.0	49.5	62.7	76.7	91.2	106.4
$3+\sqrt{6}$	26.1	38.1	51.0	64.5	78.7	93.5	108.9	124.8
4	19.3	30.5	42.6	55.5	69.2	83.5	98.4	113.9
$4+\sqrt{6}$	31.1	43.6	56.9	70.9	85.6	100.6	116.2	
5	23.9	35.7	48.3	61.9	75.7	90.4	105.6	
$5+\sqrt{6}$	36.3	49.3	63.0	77.3	92.2	107.7	123.7	

$$E(n,\ell) \simeq \left(\tfrac{1}{2}|v|\right)^{2/(v+2)} (\ell+1/2)^{2v/(v+2)} \left\{ \frac{v+2}{v} + \frac{1}{\ell+1/2}(2n+1) \right.$$

$$\sqrt{v+2} + \frac{v-2}{24(\ell+1/2)^2}\left[(2n^2+2n+1)(11-v) + \frac{2}{3}(v-8)\right] + ... \left. \right\}. \quad (2.82)$$

From this expression one gets

$$E(n,\ell+\sqrt{2+v}) - E(n+1,\ell) \simeq \left(\tfrac{1}{2}|v|\right)^{2/(v+2)}$$

$$\times \left(\ell+\frac{1}{2}\right)^{-4/(v+2)} \frac{1}{12}(n+1)(v-2)(v+1) + ... \quad (2.83)$$

This again supports the conjecture that

$$E(n,\ell+\sqrt{2+v}) > E(n+1,\ell) \quad \text{for } v > 2 \text{ (or } v < -1),$$

but

$$E(n,\ell+\sqrt{2+v}) < E(n+1,\ell) \quad \text{for } -1 < v \le 2. \quad (2.84)$$

Third argument:
A numerical evaluation of energy levels — for example, for the potential $V(r) = r^4$ — gives the values for energy levels listed in Table 5. It is seen that all values calculated for $0 \le n \le 7$ and $0 \le \ell \le 5$ support the

conjecture that $E(n, \ell) < E(n - 1, \ell + \sqrt{6})$; no violation has been found. A similar test has been made for $V = \ln r$.

As is seen for $0 \le n \le 7$ and $0 \le \ell \le 5$, we find $E(n-1, \ell + \sqrt{6}) > E(n, \ell)$. Furthermore, $(E(n-, \ell + \sqrt{6}) - E(n, \ell))(E(n, \ell) - E(n - 1, \ell))^{-1}$ does not exceed 15% for $n \le 6$.

A more favourable case is $V = r^5$. In this the effective potential

$$V = \frac{\ell(\ell + 1)}{r^2} + r^5 \tag{2.85}$$

has the property that, at $V'(r) = 0$, one has also $V'''(r) = 0$. This makes it possible to squeeze V between convenient upper and lower bounds. Specifically, after a rescaling to

$$W(\rho) = \frac{5}{\rho^2} + 2\rho^5,$$

we can prove the following chain of inequalities:

$$35(\rho-1)^2 + 21(\rho-1)^4 < \frac{5}{\rho^2} + 2\rho^5 - 7 < 35 \left(\frac{2}{\pi}\right)^2 \left[\frac{1}{\sin^2(\pi\rho/2)} - 1\right]. \tag{2.86}$$

The l.h.s. is a harmonic oscillator perturbed by a quartic term, for which, according to Loeffel, Martin, Simon and Wightman [81] diagonal Padé approximants give a lower bound. Specifically, for the Hamiltonian

$$-\frac{d^2}{dz^2} + z^2 + \lambda z^4$$

one has a lower bound,

$$E_0 > \frac{1 + \frac{5}{2}\lambda}{1 + \frac{7}{4}\lambda}.$$

The r.h.s. of (2.86) is a soluble potential according to Flügge's book on practical quantum mechanics [82]: If

$$V = \alpha^2 \left(\frac{\chi(\chi - 1)}{\sin^2 \alpha x} + \frac{\lambda(\lambda - 1)}{\cos^2 \alpha x}\right), \tag{2.87}$$

the energy levels are given by

$$E = \alpha^2 [\chi + \lambda + 2n]^2. \tag{2.88}$$

In this way one gets an explicit analytic upper bound on $E(1, \ell)$ and a lower bound on $E(0, \ell + \sqrt{7})$. For $\ell = 20$ we get

$$E(n = 0, \ \ell = 20 + \sqrt{7}) > 175.08779$$
$$E(n = 1, \ \ell = 20) < 175.07415,$$

which means that the inequality holds. A numerical calculation by Richard yields respectively 175.2 and 174.5, which means that the first bound is very good [83].

Beyond $\ell = 20$ it is possible to prove the inequality analytically. For $\ell < 20$, a numerical calculation is needed. This has been done, and for instance,

$$\begin{cases} E(n = 0, \ell = \sqrt{7}) = 15.31 \\ E(n = 1, \ell = 0) \quad = 13.43 \, . \end{cases}$$

If one accepts that programs of numerical integration of the Schrödinger equation are sufficiently accurate, this constitutes a proof for $n = 0, v = 5$.

Fourth argument:
A special case of the first part of our conjecture with $n = 1$ and $\ell = 0$ is included in the following theorem, where only monotonicity is required.

Theorem:
For a monotonous potential, $E(1, 0) < E(0, 6)$ holds — in fact we can show that $E(1, 0) < E(0, 2.5)$.

Proof:
For a monotonous potential V we can take as an upper bound a potential $\tilde{V}(r)$, being a constant $V(R)$ for $r < R$ and infinite for $r > R$. This gives an upper bound on $E(1, 0) < 4\pi^2/R^2 + V(R)$ for all R. On the other hand, the ground-state energy for angular momentum ℓ is always bounded from below by

$$E_{0,\ell} > \inf_R \left[V(R) + \frac{\ell(\ell + 1)}{R^2} \right] . \tag{2.89}$$

Taking $\ell = 6$ gives a chain of inequalities and proves the assertion. The improvement mentioned (replacing 6 by 2.5) needs a more refined argument. ∎

Conclusion — applications

Again there are at least three important applications of our results. Our first motivation came from quarkonium physics, but the results can also be applied to muonic atoms and to atomic physics, as we have noted already in this section. In the first two cases, both conditions

$$\Delta V(r) > 0 \, , \qquad D_2 V(r) < 0 \, , \tag{2.90}$$

are satisfied. In quarkonium physics all proposed potentials fulfil conditions (2.90), as can be verified by explicit calculations. Let us also note that the standard potential

$$V(r) = -\frac{\alpha(r)}{r} + kr, \quad \alpha(r)_{\text{small } r} \sim -1/\ln r \, , \tag{2.91}$$

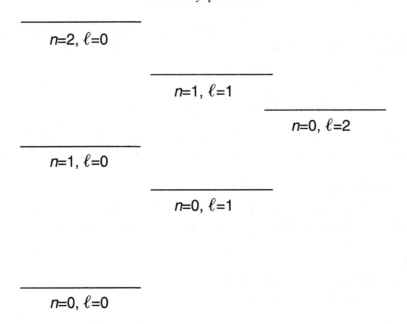

Fig. 2.3.   Ordering of levels for potentials fulfilling $\Delta V > 0$ and $D_2 V < 0$.

with a running coupling constant modifying the Coulomb-like behaviour at short distances fulfils (2.90). For small r, $\Delta V > 0$ is nothing but the antiscreening typical of QCD. In order to check (2.90) for the muonic case we consider the charge density ρ generated by V: $\rho = \Delta V$. A straightforward calculation, using the 'running elastic force constant' $K(r)$ and Newton's theorem

$$K(r_0) = r_0^{-3} \int_0^{r_0} dr\, r^2 \rho(r)\,, \qquad (2.92)$$

yields

$$r_0^2(D_2 V)(r_0) = r_0 K'(r_0) = \rho(r_0) - \langle \rho \rangle_{r \le r_0} = \frac{3}{r_0^3} \int_0^{r_0} dr\, r^2[\rho(r_0) - \rho(r)]\,, \quad (2.93)$$

where we denote by $\langle \rho \rangle_{r \le r_0}$ the mean value of $\rho(r)$ in the ball of radius r_0. A decreasing charge density thus generates a potential with $D_2 V \le 0$.

Thus for both quarkonium systems and muonic atoms (here we use the experimental fact that for most nuclei the charge density is decreasing from the centre of the nucleus) the energy level ordering is between that of the hydrogen atom and the harmonic oscillator, in the sense that (see Figure 2.3)

$$E(n+1, \ell) \ge E(n, \ell+1) \ge E(n+1, \ell)\,. \qquad (2.94)$$

The third application to atomic physics was discussed in Section 1.1. Let us note here in addition that for excited-state energy levels of the neutral alkali-metal atoms Sternheimer [84] introduced a new ordering scheme. Motivated by measured energy level values he observed near-degeneracy of levels when $n + 2\ell$ is introduced as a new quantum number. This situation corresponds to $\alpha = 1/2$. From Theorem (2.47) we observe that there exists a potential giving exact degeneracy for such levels at one energy.

Remark:

The previous results were obtained for potentials which fulfilled conditions like (2.39) where the second order differential operator D_α, Eq. (2.40), appeared. It turns out that in two further problems, one concerning the relationship between moments [85] and the other constraints on the total and kinetic energy of ground states [86], similar conditions will be used. For convenience, we define two classes of potentials:

Class A: For all $r > 0$, <u>one</u> of the following conditions holds

$$(i) \qquad D_\alpha V(r) > 0, \ 1 \le \alpha \le 2, \ V(r) \ge 0,$$

$$(ii) \qquad D_\alpha V(r) > 0, \ \alpha < 1, \ V(r) \le 0, \tag{2.95}$$

$$(iii) \qquad \left(r^2 \frac{d^2}{dr^2} + (3 - 2\alpha)\frac{d}{dr} \right) V(r) \ge 0, \ 1 < \alpha < 2, \ \frac{dV}{dr} \ge 0.$$

Class B: For all $r > 0$, <u>one</u> of the following conditions holds

$$(i) \qquad D_\alpha V(r) \le 0, \ \alpha > 2 \text{ or } < 1, \ V(r) \ge 0,$$

$$(ii) \qquad D_\alpha V(r) \le 0, \ 1 \le \alpha \le 2, \ V(r) \le 0, \tag{2.96}$$

$$(iii) \qquad \left(r^2 \frac{d^2}{dr^2} + (3 - 2\alpha)\frac{d}{dr} \right) V(r) \le 0, \ \alpha \ge 2.$$

Splitting of Landau levels in non-constant magnetic fields

It has been recognized that the degeneracy of the so-called Landau levels, the levels of a charged particle confined to a plane subjected to a constant magnetic field, play an essential role in the quantum Hall effect [87]. We thought that it might be of some interest to use the 'technology' developed in the case of central potentials to study how the degeneracy is removed when the magnetic field is not constant [88], and to consider, first, the simplest case — the one in which the magnetic field has cylindrical symmetry around an axis perpendicular to the plane.

So we consider the magnetic Schrödinger operator

$$H_{\vec{A}} = (-i\vec{\nabla} + \vec{A})^2 \tag{2.97}$$

in two space dimensions.

Fixing the gauge, we suppose that $\vec{A}$ is of the form

$$\vec{A}(x, y) = a(r) \begin{pmatrix} -y \\ +x \end{pmatrix}, \quad r = \sqrt{x^2 + y^2}. \tag{2.98}$$

Then the magnetic field B depends only on r and is given by

$$B(r) = \partial_x A_y - \partial_y A_x = 2a(r) + r\frac{da(r)}{dr}. \tag{2.99}$$

Since the angular momentum is conserved, $H_{\vec{A}}$ splits into an infinite family of radial operators which reads, in its reduced form,

$$H_m = -\frac{d^2}{dr^2} + \frac{m^2 - \frac{1}{4}}{r^2} + a^2(r)r^2 - 2ma(r). \tag{2.100}$$

In the case of a constant magnetic field, B (i.e., $a = B/2$ in our gauge), the energies are given by

$$E(n, m) = B(2n + 1), \tag{2.101}$$

where n denotes the number of nodes of the corresponding reduced radial wave function. Therefore, all energy levels are infinitely degenerate (in m).

For general $B(r)$ the Schrödinger operator $H_{\vec{A}}$ may have rather surprising spectral properties. We illustrate this fact by an example given in Ref. [89]: We take $a(r) = k(1 + r)^{-\gamma}, \gamma > 0$. Hence $B(r) = 2k(1 + r)^{-\gamma} - k\gamma r(1 + r)^{-\gamma-1}$ and $B(r) \to 0$ as $r \to \infty$. Therefore, the essential spectrum of $H_{\vec{A}}$ is equal to $\mathbf{R}_0^+$. There are the following three cases:

$\gamma < 1$: Then the potential term in each H_m goes to infinity as $r \to \infty$. So H_m and therefore $H_{\vec{A}}$ has a pure point spectrum (i.e., only eigenvalues). On the other hand, the spectrum equals $\mathbf{R}_0^+$. Therefore, the point spectrum is dense.

$\gamma = 1$: Then the potential shifted by $-k^2$ is a long-range potential going to zero as r tends to infinity. Hence H_m (and therefore $H_{\vec{A}}$) has an (absolutely) continuous spectrum in (k^2, ∞) and a dense, pure-point spectrum in $(0, k^2)$.

$\gamma > 1$: In this case the potential term is a short-range potential, and therefore there is only a continuous spectrum.

Our main results are the following:

Theorem:

Let $E_{m,n}(\lambda)$ denote the eigenvalues of the two-dimensional magnetic Schrödinger operator $H_{\vec{A}}$ corresponding to a magnetic field, orthogonal to the plane, of strength $B(r) = B_0 + \lambda B_1(r), \lambda \geq 0$, where r is the two-dimensional distance.

Then we have, for all $n \geq 0$ and $m \geq 0$,

$$\lim_{\lambda \to 0} \frac{1}{\lambda} \left[E_{m+1,n}(\lambda) - E_{m,n}(\lambda) \right] \gtrless 0 \quad \text{if} \quad \frac{dB}{dr} \gtrless 0. \tag{2.102}$$

Our second result holds beyond first-order perturbation theory, but only for purely angular excitations.

Theorem:

Consider the two-dimensional magnetic Schrödinger operator for the field $B(r)$. Suppose that the field $B(r)$ is positive for large r. Then

$$E_{m+1,0} \gtrless E_{m,0} \quad \text{if} \quad \frac{dB}{dr} \gtrless 0. \tag{2.103}$$

From the physical point of view the ordering result is quite obvious.

The distance of an eigenstate from the origin increases with the angular momentum m. Thus, if the field B grows (or decreases respectively) with the distance, the eigenstates feel a stronger (weaker) field and therefore the energies should increase (decrease) as a function of m within a Landau level.

For the proof of Theorem (2.102) we shall use the supersymmetric structure of the constant-field Schrödinger operator, i.e., H_m is factorizable and there are ladder operators acting between the H_ms.

For Theorem (2.103) we use the supersymmetry of the operators $H_{\hat{A}} - B(r)$ and a comparison theorem for the purely angular excitations of $H_{\hat{A}}$.

Proof: Theorem (2.102)

First of all, we present a suitable factorization of the reduced Hamiltonians H_m in the case of a constant magnetic field. We have

$$H_m = -\frac{d^2}{dr^2} + \frac{m^2 - \frac{1}{4}}{r^2} + a_0^2 r^2 - 2ma_0, \tag{2.104}$$

where $a_0 = B_0/2$. Its spectrum is $(4a_0(n+1/2), n \epsilon N^0)$. We define the ladder operator and its conjugate

$$A_m = -\frac{d}{dr} + \frac{m + \frac{1}{2}}{r} + a_0 r, \qquad A_m^+ = \frac{d}{dr} + \frac{m + \frac{1}{2}}{r} + a_0 r. \tag{2.105}$$

We have

$$A_m^+ A_m = H_m + 4a_0 \left(m + \frac{1}{2} \right), \qquad A_m A_m^+ = H_{m+1} + 4a_0 \left(m + \frac{1}{2} \right) \tag{2.106}$$

and the intertwining relation

$$H_m A_m^+ = A_m^+ H_{m+1} \,. \tag{2.107}$$

If we denote by $u_{n,m}$, the reduced eigenfunctions we find

$$A_m u_{n,m} = c_{nm} u_{n,m+1} \,, \qquad A_m^+ u_{n,m+1} = c_{nm} u_{n,m} \,, \tag{2.108}$$

with $c_{nm}^2 = 4a_0(n + m + 1)$.

From the ladder operators we deduce the following useful sum rules. Let W be a differentiable spherically symmetric function. Then, denoting by $\langle .|. \rangle$ the L^2 inner product with respect to dr,

$$c_{nm}\langle u_{n,m}|W|u_{n,m+1}\rangle = \langle u_{n,m}|W|A_m u_{n,m}\rangle = \langle A_m^+ u_{n,m+1}|W|u_{n,m+1}\rangle \tag{2.109}$$

and therefore

$$\left\langle u_{n,m} \left| \frac{W'}{2} + \left(\frac{m + \frac{1}{2}}{r} + a_0 r \right) W \right| u_{n,m} \right\rangle$$
$$= \left\langle u_{n,m+1} \left| -\frac{W'}{2} + \left(\frac{m + \frac{1}{2}}{r} + a_0 r \right) W \right| u_{n,m+1} \right\rangle . \tag{2.110}$$

In the general case we write the radial part of the vector field as $a_0 + \lambda a(r)$. In the first-order approximation (in λ) the energies are given by

$$E_{n,m}(\lambda) = 4a_0 \left(n + \frac{1}{2} \right) + 2\lambda \langle u_{n,m}|a_0 a(r)r^2 - ma(r)|u_{n,m}\rangle \,. \tag{2.111}$$

Therefore, for the splitting we have

$$\Delta_{n,m}^{n,m+1} \equiv \frac{1}{2\lambda}\left[E_{n,m+1}(\lambda) - E_{n,m}(\lambda) \right]$$
$$= \left\langle u_{n,m+1} \left| \left[a_0 r^2 - \left(m + \frac{1}{2} \right) \right] a(r) + \frac{a(r)}{2} \right| u_{n,m+1} \right\rangle \tag{2.112}$$
$$- \left\langle u_{n,m} \left| \left[a_0 r^2 - \left(m + \frac{1}{2} \right) \right] a(r) - \frac{a(r)}{2} \right| u_{n,m} \right\rangle .$$

Applying the sum rule with $W' = a(r)$ we find

$$\Delta_{n,m}^{n,m+1} = \left\langle u_{n,m+1} \left| \left(a_0 r^2 - m - \frac{1}{2} \right) a(r) - \left(\frac{m + \frac{1}{2}}{r} + a_0 r \right) W \right| u_{n,m+1} \right\rangle$$
$$\tag{2.113}$$
$$- \left\langle u_{n,m} \left| \left(a_0 r^2 - m - \frac{1}{2} \right) a(r) - \left(\frac{m + \frac{1}{2}}{r} + a_0 r \right) W \right| u_{n,m} \right\rangle .$$

Again using the ladder operators we obtain

$$\Delta_{n,m}^{n,m+1} = \frac{1}{c_{nm}}\langle u_{n,m}|U'|u_{n,m+1}\rangle \,, \tag{2.114}$$

where

$$U = \left(a_0 r^2 - m - \frac{1}{2}\right) a(r) - W(r) \left(\frac{m + \frac{1}{2}}{r} + a_0 r\right). \tag{2.115}$$

Since

$$U' = \frac{d}{dr} U = \left(a_0 - \frac{m + \frac{1}{2}}{r^2}\right) \left(r^2 \frac{da(r)}{dr} + ra(r) - W(r)\right) \tag{2.116}$$

and

$$\frac{d}{dr} \left(r^2 \frac{da(r)}{dr} + ra(r) - W(r)\right) = r^2 \frac{d^2}{dr^2} a(r) + 3r \frac{da(r)}{dr} = r \frac{dB(r)}{dr}, \tag{2.117}$$

we have, after integrating by parts,

$$\Delta_{n,m}^{n,m+1} = \frac{1}{c_{nm}} \int r \frac{dB(r)}{dr} \int_r^\infty \left(a_0 - \frac{m + \frac{1}{2}}{s^2}\right) u_{n,m}(s) u_{n,m+1}(s) ds \, dr. \tag{2.118}$$

Now we can show that

$$F(r) \equiv \frac{1}{c_{nm}} \int_r^\infty \left(a_0 - \frac{m + \frac{1}{2}}{s^2}\right) u_{n,m}(s) u_{n,m+1}(s) ds \tag{2.119}$$

is strictly positive for $r > 0$. Using $u_{n,m+1} = (1/c_{nm}) A_m u_{n,m}$ we find using integration by parts

$$F(r) = \frac{1}{2} u_{n,m}^2(r) \left(a_0 - \frac{m + \frac{1}{2}}{r^2}\right) + \int_r^\infty \left(a_0^2 s - \frac{m^2 - \frac{1}{4}}{s^3}\right) u_{n,m}^2(s) ds. \tag{2.120}$$

On the other hand, multiplying the eigenvalue equation for $u_{n,m}$ by (d/dr) $u_{n,m}$ and integrating from r to infinity we get

$$\left(\frac{d}{dr} u_{n,m}(r)\right)^2 - \frac{m^2 - \frac{1}{4}}{r^2} u_{n,m}^2(r) - a_0^2 r^2 u_{n,m}^2(r)$$

$$+ 2a_0(2n + m + 1) u_{n,m}^2(r) = 2 \int_r^\infty \left(a_0^2 s - \frac{m^2 - \frac{1}{4}}{s^3}\right) u_{n,m}^2(s) ds. \tag{2.121}$$

Hence

$$F(r) = \frac{1}{2} \left[\left(\frac{d}{dr} u_{n,m}(r)\right)^2 - \left(\frac{(m + \frac{1}{2})^2}{r^2} - 2a_0 \left(2n + m + \frac{3}{2}\right) + a_0^2 r^2\right) u_{n,m}^2(r)\right].$$

$$\tag{2.122}$$

Now $F(0) = 0$ and $\lim_{r \to \infty} F(r) = 0$. At critical points of $F(r)$ we have the following possibilities.

(i) If $u_{n,m}(r) = 0$ then
$((d/dr) u_{n,m}(r))^2 > 0$, hence $F(r) > 0$.

(ii) If $a_0 r^2 = m + 1/2$ then
$$F(r) = ((d/dr)u_{n,m}(r))^2 + a_0(4m + 2)u_{n,m}^2(r) > 0.$$

(iii) If $(d/dr)u_{n,m}(r) = (a_0 r + (m + 1/2)/r)u_{n,m}(r)$ then
$$F(r) = (4n + 2)a_0 u_{n,m}^2(r) > 0.$$

Hence Theorem (2.102) is proved. ∎

Proof: Theorem (2.103)

We study the one-parameter family of Hamiltonians

$$H_{\vec{A}}(\lambda) = H_{\vec{A}} - \lambda B(r), \quad \text{with} \quad B(r) \geq 0 \text{ for large } r. \tag{2.123}$$

For $\lambda = 0$ we have the operator describing the motion of a spinless particle in the magnetic field $B(r)$, while for $\lambda = 1$ the Hamiltonian $H_{\vec{A}}(1)$ is the projection onto the spin-down component of the Hamiltonian describing the motion of a spin-1/2 particle in the field $B(r)$. This Hamiltonian, however, has an infinitely degenerate ground state of energy $E = 0$ [89].

In fact, considering the family of reduced Hamiltonians

$$H_m(\lambda) = H_m(0) - \lambda B(r) \tag{2.124}$$

with

$$H_m(0) = -\frac{d^2}{dr^2} + \frac{m^2 - \frac{1}{4}}{r^2} + a^2(r)r^2 - 2ma(r) \tag{2.125}$$

and denoting the nodeless, reduced eigenfunctions by $u_m(\lambda)$, i.e.,

$$H_m(\lambda)u_m(\lambda) = E_m(\lambda)u_m(\lambda), \tag{2.126}$$

we have $E_m(1) = 0$ for all $m \geq 0$ and

$$u_m(1, r) = c_m r^{m + \frac{1}{2}} \exp\left(-\int^r sa(s)ds\right). \tag{2.127}$$

By the Feynman–Hellmann theorem the derivative of $E_m(\lambda)$ with respect to λ equals the expectation value of the magnetic field, so that we find for the splitting of two eigenvalues

$$\frac{d}{d\lambda}[E_{m+1}(\lambda) - E_m(\lambda)] = \langle u_m(\lambda)|B|u_m(\lambda)\rangle - \langle u_{m+1}(\lambda)|B|u_{m+1}(\lambda)\rangle$$

$$= \int_0^\infty \left(\frac{d}{dr}B(r)\right)I(r, \lambda)dr, \tag{2.128}$$

with

$$I(r, \lambda) = \int_0^r \left[u_{m+1}^2(\lambda, s) - u_m^2(\lambda, s)\right]ds. \tag{2.129}$$

We have to consider the cases $dB/dr > 0$ and $dB/dr < 0$ separately.

$\underline{dB/dr > 0}$

We want to show $I(r, \lambda) < 0$. To do so we analyse the Wronskian of u_{m+1} and u_m

$$w(r) = u_m(\lambda, r)\frac{d}{dr}u_{m+1}(\lambda, r) - u_{m+1}(\lambda, r)\frac{d}{dr}u_m(\lambda, r) ; \qquad (2.130)$$

$w(r)$ vanishes both at zero and at infinity and

$$\frac{d}{dr}w(r) = u_m(\lambda, r)u_{m+1}(\lambda, r)\left[\frac{2m+1}{r^2} - 2a(r) + E_m(\lambda) - E_{m+1}(\lambda)\right] . \qquad (2.131)$$

If we denote the square bracket by $h(r)$ we have

$$\frac{dh(r)}{dr} = -\frac{2}{r^3}\left(2m + 1 + r^3\frac{da(r)}{dr}\right) . \qquad (2.132)$$

For fields $B(r)$, finite at the origin (less singular than a two-dimensional delta-function is also sufficient), $r^3(da(r)/dr)$ equals zero at the origin. Since

$$\frac{dB}{dr} = \frac{1}{r^2}\frac{d}{dr}\left(r^3\frac{da(r)}{dr}\right) > 0 , \qquad (2.133)$$

the square bracket is a strictly decreasing function and therefore the positivity of $w(r)$ implies that $u_{m+1}(\lambda, r)/u_m(\lambda, r)$ is strictly increasing. Since $I(r, \lambda)$ vanishes at zero and at infinity we have $I(r, \lambda) < 0$. Hence:

$$\frac{d}{d\lambda}[E_{m+1}(\lambda) - E_m(\lambda)] < 0 . \qquad (2.134)$$

Integrating from 0 to 1 with respect to λ we find

$$E_{m+1}(0) - E_m(0) > E_{m+1}(1) - E_m(1) = 0 . \qquad (2.135)$$

$\underline{dB/dr < 0}$

We prove the ordering result by contradiction. Again we analyse the Wronskian $w(r)$ of u_{m+1} and u_m. We write its derivative as follows using $B(r) = 2a(r) + r(da(r)/dr)$

$$\frac{d}{dr}w(r) = \frac{u_m(\lambda, r)u_{m+1}(\lambda, r)}{r^2}$$

$$\times \left[2m + 1 - B(r) \cdot r^2 + r^3\frac{da(r)}{dr} + r^2\left((E_m(\lambda) - E_{m+1}(\lambda))\right)\right] . \qquad (2.136)$$

Denoting the square bracket by $k(r)$ we have

$$\frac{dk(r)}{dr} = 2r(E_m(\lambda) - E_{m+1}(\lambda) - B(r)) . \qquad (2.137)$$

Now we prove the result by contradiction. For this we note that $I(r, 1) < 0$, as can easily be checked. Hence:

$$\frac{d}{d\lambda} [E_{m+1}(\lambda) - E_m(\lambda)] \,|_{\lambda=1} > 0 \,. \qquad (2.138)$$

Thus, either $E_{m+1}(\lambda) < E_m(\lambda)$ for all $\lambda \epsilon [0, 1)$, or there is a first point λ^* (starting from 1) such that $E_{m+1}(\lambda^*) = E_m(\lambda^*)$. But then $dk(r)/dr = -2rB(r) \leq 0$ since $B(r)$ is non-negative by assumption.

As before, we conclude that $I(r, \lambda^*) < 0$. Hence:

$$\frac{d}{d\lambda} [E_{m+1}(\lambda) - E_m(\lambda)] \,|_{\lambda=\lambda^*} > 0 \,, \qquad (2.139)$$

which yields the desired contradiction.

Remark:

Our proof does not work for excited states, since the lemma does not hold for states which have nodes.

Generalizations of the above results to Hamiltonians with an additional scalar potential have been worked out by one of us (H. G.) and Stubbe [88].

2.3   Spacing of energy levels

We turn now to the problem of the spacing of the energy levels, and, first of all, to the spacing between the angular excitations — i.e., the states with $n = 0$ (zero node) and ℓ arbitrary. For this we need first to present the remarkable discovery of Common [90] concerning the moments of the wave functions of these angular excitations:

$$\langle r^\nu \rangle_\ell = \int_0^\infty u_{0,\ell}^2(r) r^\nu dr \,, \qquad (2.140)$$

which follows from lemma (2.28). This says that for a given sign of the Laplacian of the potential the reduced wave function divided by $r^{\ell+1}$ is logarithmically concave ($\Delta V > 0$) or convex ($\Delta V < 0$). We shall not follow the original derivation of Common [90], which used integration by parts, and instead shall present the following somewhat stronger result, obtained by Common, Martin and Stubbe [91].

Theorem:

If the potential has a positive (negative) Laplacian outside the origin, the quantity

$$M_\ell(\nu) = \frac{\langle r^\nu \rangle_\ell}{\Gamma(2\ell + \nu + 3)} \,, \qquad (2.141)$$

where the numerator is the expectation value of r^{v} in the ground state of angular momentum ℓ, is respectively logarithmically concave or convex — i.e., $\log(M_{\ell}(v))$ is concave or convex. Before giving the proof, let us indicate that in Common's original result the moments jumped by one unit. For instance,

$$\langle r^{-1} \rangle_{\ell=0} \langle r \rangle_{\ell=0} < \frac{3}{2} |\langle 1 \rangle_{\ell=0}|^2 \qquad (2.142)$$

for $\Delta V > 0$.

Another important remark is that Common's original inequalities, as well as the more general ones, lead to inequalities of the type of the Schwarz inequalities for $\Delta V < 0$, while they give inequalities going in the opposite direction, such as the example given above, for $\Delta V > 0$.

Now we start with the proof. First of all we shall use the function

$$\phi(r) = \left(\frac{u_{\ell}(r)}{r^{\ell+1}} \right)^2 , \qquad (2.143)$$

which is logarithmically concave (convex) if the Laplacian of the potential is negative (positive) for all $r > 0$. Then the problem is to study the logarithmic concavity of

$$N(v) = \frac{1}{\Gamma(v+1)} \int \phi(r) r^{v} dr . \qquad (2.144)$$

For this purpose we consider the quantity

$$N(v) - 2xN(v+\epsilon) + x^2 N(v+2\epsilon)$$

$$= \int \phi(r) \left[\frac{r^{v}}{\Gamma(v+1)} - 2x \frac{r^{v+\epsilon}}{\Gamma(v+1+\epsilon)} + x^2 \frac{r^{v+2\epsilon}}{\Gamma(v+1+2\epsilon)} \right] dr. \quad (2.145)$$

The square bracket in Eq. (2.145) has two zeros in $\mathbf{R}^{+}$ because it is a second degree polynomial in xr^{ϵ} with a positive discriminant as a consequence of the logarithmic convexity of the gamma-functions. Let $0 < y_1 < y_2$ be the two solutions of the equation

$$\frac{1}{\Gamma(v+1)} - \frac{2y}{\Gamma(v+1+\epsilon)} + \frac{y^2}{\Gamma(v+1+2\epsilon)} = 0 , \qquad (2.146)$$

then the square bracket in (2.145) vanishes at

$$r_1 = \left(\frac{y_1}{x} \right)^{\frac{1}{\epsilon}} \quad r_2 = \left(\frac{y_2}{x} \right)^{\frac{1}{\epsilon}} . \qquad (2.147)$$

Now we construct the expression

$$X = \int \left[\phi(r) - Ae^{-Br}\right] \left[\frac{r^v}{\Gamma(v+1)} - \frac{2xr^{v+\epsilon}}{\Gamma(v+1+\epsilon)} + \frac{x^2 r^{v+2\epsilon}}{\Gamma(v+1+2\epsilon)}\right] dr,$$

(2.148)

where A and B are adjusted in such a way that the first square bracket vanishes at r_1 and r_2. Because of the logarithmic convexity or concavity of ϕ, this is always possible, and the first square bracket has no other zero. Hence we get

$$N(v) - 2xN(v+\epsilon) + x^2 N(v+2\epsilon) - A\left(1 - \frac{x}{B^\epsilon}\right)^2 \gtrless 0 \qquad (2.149)$$

if $\Delta V \gtrless 0$.

The simplest case is $\Delta V < 0$. Then, from (2.149) and from the positivity of A it follows that for any x

$$N(v) - 2xN(v+\epsilon) + x^2 N(v+2\epsilon) > 0,$$

and this implies

$$(N(v+\epsilon))^2 < N(v)N(v+2\epsilon),$$

which demonstrates the logarithmic convexity of N, and hence of $M_\ell(v)$.

For $\Delta V > 0$, one must find x and B such that $x = B^\epsilon$. Then, the combination of Ns in (2.149) is negative for some x, and this implies that its discriminant is positive. But this can be done only by a fixed point argument that we now sketch. The key point is that we have to impose $\phi(r_2)/\phi(r_1) = \exp(-B(r_2 - r_1))$ with $R = r_2/r_1 = (y_2/y_1)^{1/\epsilon}$, y_1 and y_2 being given uniquely by Eq. (2.146). Furthermore, $x = B^\epsilon = y_1/r_1^\epsilon$.

Hence, we have to show the existence of a solution of $\phi(Rr_1)/\phi(r_1) = \exp(-y_1^{1/\epsilon}(R-1))$; the l.h.s. goes to unity for $r_1 \to 0$ and to $-\infty$ for $r_1 \to \infty$ because of the logarithmic concavity of ϕ. There is therefore a solution and the second part of the theorem is proved.

We leave it as an exercise for the reader to generalize these results to mixed moments, i.e., $\int u_{0,\ell} u_{0,\ell'} r^v dr$.

It remains to generalize these results to the case where, instead of having a Laplacian of a given sign, the potential belongs to one of the sets A and B previously introduced in Eqs. (2.95) and (2.96), which are obtained by a change of variables. We shall limit ourselves here to the special case, where $(d/dr)(1/r)(dV/dr)$ has a given sign, i.e., where V is concave or convex as a function of r^2. Then all one has to do is to change variables $z = r^2$, and the theorem becomes:

Theorem:

If V is convex (concave) in r^2

$$\log \frac{\int u_{0,\ell}^2 r^\nu \, dr}{\Gamma\left((\nu + 2\ell + 3)/2\right)} \quad \text{is concave (convex) in } \nu . \tag{2.150}$$

Now, we have all that we need to study the spacing of energy levels. There are, of course, other consequences of Common's result — for instance, inequalities on the kinetic energy and the wave function — but these will be considered later.

The first problem we attack is the study of the concavity or convexity of $E(\ell)$, the energy of the ground state, characterized by $n = 0$ and ℓ, as a function of ℓ. For a harmonic oscillator potential we know that $E(\ell)$ is a linear function of ℓ. For a general potential we know first of all that $E(\ell)$ is increasing because by the Feynman–Hellmann theorem

$$\frac{dE}{d\ell}(\ell) = (2\ell + 1) \int_0^\infty \frac{u_\ell^2}{r^2} dr , \tag{2.151}$$

and we also know that it increases more slowly than $\ell(\ell + 1)$, since, according to a general theorem indicated in Section 2.1, the energy of the ground state is a concave function of any parameter entering linearly in the Hamiltonian. For an infinite square well $E(\ell)$ indeed behaves like $\ell(\ell + 1)$ for large ℓ.

We return now to the comparison with the harmonic oscillator. First we consider the case of a potential <u>concave</u> in r^2 — i.e., such that $(d/dr)(1/r)(dV/dr) < 0$. Naturally, the angular momentum can be taken as a continuous variable, and we have to prove that the quantity

$$2E(\ell) - E(\ell + \delta) - E(\ell - \delta)$$

is positive for arbitrary δ to establish that the energy is a <u>concave</u> function of ℓ. An upper bound on $E(\ell - \delta)$ and $E(\ell + \delta)$ can easily be obtained by using the trial functions $r^{-\delta} u_\ell(r)$ and $r^{+\delta} u_\ell(r)$, respectively (notice that these trial functions become <u>exact</u> in the harmonic oscillator case).

Elementary algebraic manipulations, using the Schrödinger equation for u_ℓ, lead to

$$2E(\ell) - E(\ell - \delta) - E(\ell - \delta) >$$

$$\delta \left[(2\ell + 1 - 2\delta) \frac{\int u^2 r^{-2-2\delta} \, dr}{\int u^2 r^{-2\delta} \, dr} - (2\ell + 1 + 2\delta) \frac{\int u^2 r^{-2+2\delta} \, dr}{\int u^2 r^{2\delta} \, dr} \right] . \tag{2.152}$$

From the concavity of $\log \langle r^\nu \rangle / \Gamma((2\ell + 3 + \nu)/2)$, it is easy to see that the r.h.s. of (2.152) is positive. Therefore, we have

Theorem:

If $(d/dr)(1/r)(dV/dr) < 0$ for all $r > 0$, $E(\ell)$ is concave in ℓ.

Its converse is also true, but the demonstration is 'trickier', and is given in Appendix B. At any rate, we have obtained [48] the theorem:

$$\left.\begin{array}{l} \dfrac{d}{dr}\dfrac{1}{r}\dfrac{dV}{dr} \gtrless 0 \;\; \forall r > 0 \\[4mm] \Rightarrow \dfrac{d^2 E}{d\ell^2} \gtrless 0 \end{array}\right\}. \tag{2.153}$$

The borderline case is naturally the harmonic oscillator for which E is a linear function of ℓ.

Subsequently, the analogue theorem has been obtained for the case of potentials with a Laplacian of a given sign [50]:

Theorem:

If

$$\left.\begin{array}{l} \dfrac{d}{dr} r^2 \dfrac{dV}{dr} \gtrless 0 \;\; \forall r > 0 \\[4mm] \Rightarrow \dfrac{d^2 E}{d\ell^2} \gtrless - \dfrac{3}{\ell+1}\dfrac{dE}{d\ell} \end{array}\right\}. \tag{2.154}$$

Although we do not want to give the proofs of these two theorems in the text let us indicate the essence of the proofs. In both cases extensive use is made of the Chebyshev inequality:

If f and g are both decreasing or both non-increasing and h is non-negative:

$$\int fh dx \int gh dx \leq \int fgh dx \int h dx. \tag{2.155}$$

For (2.153) one uses

$$-\left(\dfrac{u'}{ru}\right)' - \dfrac{2(\ell+1)}{r^3} > 0 \tag{2.156}$$

and for (2.154)

$$-\left(\dfrac{u'}{u}\right)' - \dfrac{\ell+1}{r^2} > 0. \tag{2.157}$$

A weaker, but more transparent, version of the second theorem is obtained by integration. That is,

$$\dfrac{E_{\ell+1} - E_\ell}{E_\ell - E_{\ell-1}} \gtrless \dfrac{E_{\ell+1}^c - E_\ell^c}{E_\ell^c - E_{\ell-1}^c}, \tag{2.158}$$

if $\Delta V(r) \gtrless 0$ $\forall r > 0$, and where E_ℓ^c designates the Coulomb energy. Then the r.h.s. of (2.158) is

$$\frac{2\ell + 3}{2\ell + 1} \left(\frac{\ell}{\ell + 2} \right)^2 . \tag{2.159}$$

In the introduction, we have given illustrations of this inequality, with muonic atoms — for which $\Delta V(r) > 0$ — and alkaline atoms, for which, in the one-electron approximation, one has $\Delta V(r) < 0$.

Until now we have ignored spin effects, but the technology developed so far makes it possible to take them into account, in the case of spin $1/2$ particles, in a semirelativistic treatment 'à la Pauli'. Then it is known that the splitting between the states $J = \ell + 1/2$ and $J = \ell - 1/2$ is, to first order in an expansion of the energies in $(v/c)^2$, given by

$$\delta = \frac{2\ell + 1}{4m^2} \left\langle \frac{dV/dr}{r} \right\rangle , \tag{2.160}$$

where the expectation value is to be taken between Schrödinger wave functions of given orbital angular momentum and where V is the central potential entering the corresponding Dirac operator as a vector-like quantity. We shall designate $\delta(\ell)$ the splitting corresponding to a purely angular excitation (i.e., with a nodeless radial wave function).

Then, we have two theorems [50]:

$$\delta(\ell) \lessgtr \Delta(\ell) = \frac{2}{m\ell} \frac{(\ell + 2)^4}{(2\ell + 3)^2} [E(\ell + 1) - E(\ell)]^2 , \tag{2.161}$$

for $\Delta V(r) \gtrless 0$ $\forall r > 0$, and

$$\frac{\delta(\ell + 1)}{\delta(\ell)} \gtrless \frac{\ell(\ell + 1)^3}{(\ell + 2)^4} , \tag{2.162}$$

for $\Delta V(r) \gtrless 0$ $\forall r > 0$.

For their proof, the Chebyshev inequality previously mentioned in Eq. (2.155) is essential, but we shall simply refer the reader to the original reference. The most beautiful illustration of these inequalities is given by muonic atoms (see Table 6). For small Z, $\Delta(\ell)$ the upper bound almost coincides with the experimental value because the nucleus is almost point-like, while, for large Z — Lead, for instance — Δ/δ exceeds 2.

Another area where similar interesting inequalities can be obtained is that of the fine structure of the P states of quarkonium, which consist of a triplet and a singlet. This structure is completely analogous to the fine structure of positronium. Strangely enough the singlet P state of the $c\bar{c}$ system has been observed and its mass accurately measured before the singlet P state of positronium [92].

Table 6.   Fine splittings of levels of muonic atoms (δ), their upper bounds (Δ) and their ratios.

Element	Z	$\delta(1)$ keV	$\Delta(1)$ keV	$\delta(2)$ keV	$\Delta(2)$ keV	$\delta(3)$ keV	$\delta(2)/\delta(1)$ > 0.0988	$\delta(3)/\delta(2)$ > 0.211
^{56}Fe	26	4.20 ±0.09	4.40	0.47 ±0.07			0.112 ±0.02	
^{Nat}Ni	28	5.30	5.89					
^{Nat}Cu	29	6.17	6.78	1.68 ±0.6	±0.7		0.26 ±0.13	
^{68}Zn	30	7.00 ±0.6	7.84	0.56 ±0.65			0.08 ±0.012	
^{75}As	33	11.10 ±0.8	11.45	2.00 ±0.8			0.18 ±0.08	
^{89}Y	39	19.26 ±0.94	22.70	2.27 ±0.8	2.74	2.27 ±1.8	0.188 ±0.04	1.00 ±0.8
^{93}Nb	41	23.15 ±1	29.10	2.68 ±0.22			0.116 ±0.01	
^{92}Mo	42	25.50 ±0.8	29.60	2.7 ±0.4			0.106 ±0.015	
^{114}Cd	48	39.90 ±0.8	52.50	5.16 ±0.5	5.07	2.20 ±0.8	0.129 ±0.015	0.43 ±0.2
^{115}In	49	43.60 ±0.8	56.80	5.35 ±0.96			0.123 ±0.02	
^{120}Sn	50	45.70 ±0.9	60.60	5.65 ±0.9	5.98	0.95 ±0.6	0.123 ±0.02	0.43 ±0.2

There are four P states in a $q\bar{q}$ system (where q is a quark, or an electron), three triplet P states with total angular momenta $J = 0$, 1, 2 and one singlet P state. In the framework of the Breit–Fermi approximation [93], which may or may not be correct, and assuming that the central potential is made of scalar-like and vector-like parts

$$V = V_s + V_v , \qquad (2.163)$$

one finds

$$M(1^1P_1) = \tilde{M} - \frac{1}{2m^2}\left\langle \frac{1}{r}\frac{dV}{dr} \right\rangle , \qquad (2.164)$$

where

$$\tilde{M} = \frac{1}{12}\left[-5M(1^3P_2) + 27M(1^3P_1) - 10(1^3P_0) \right] \qquad (2.165)$$

(notice that $\tilde{M}$ is <u>NOT</u> the centre of gravity of the triplet P states).

One could, of course, use our previous result (2.161) to get a lower bound on $M(1^1P_1)$, since we believe that the quark–antiquark potential has a positive Laplacian. But this turns out not to be terribly constraining. We prefer to use the information that V is concave, to get

$$\left\langle \frac{V'}{r} \right\rangle \geq \frac{5}{8} m[E(2) - E(1)]^2 . \tag{2.166}$$

Here again, the Chebyshev inequality is used in the proof. For the converse inequalities

$$\left\langle \frac{V'}{r} \right\rangle \leq \frac{m}{2}[E(1) - E(0)]^2 \tag{2.167}$$

we only need $V'' < (V'/r)$, but an extra tool is the mixed sum rule already mentioned at the end of Section 2.1:

$$\int_0^\infty \frac{dV}{dr} u_1 u_0 dr = \frac{1}{2}[E(1) - E(0)]^2 \int_0^\infty r u_1 u_0 dr .$$

The final conclusion for the case of charmonium when one puts numbers is, for $m_c = 1.5$ GeV,

$$3536.2 \pm 12.2 \text{ MeV} < M(1^1P_1) < 3558.6 \pm 12.2 \text{ MeV} ,$$

where the error is a crude estimate of relativistic effects. This is in agreement with the observation of a $^1P_1$ state, six months after this prediction in Ref. [94] $M = 3.526$ GeV.

For the Upsilon system, we predict

$$9900.3 \pm 2.8 \text{ MeV} < M(1^1P_1) < 9908 \pm 2.8 \text{ MeV} .$$

We now turn to the spacing of radial excitations — i.e., excitations where the angular momentum is fixed and the number of nodes increases. A number of results have been obtained by Richard, Taxil and one of the authors (A.M.) [49]. As we shall see, these results are interesting but incomplete and some conjectures still need a completely rigorous proof. Again, the harmonic oscillator will be our starting point. Indeed the $\ell = 0$ levels of the three-dimensional harmonic oscillator are equally spaced:

$$2E(n, \ell = 0) = E(n+1, \ell = 0) + E(n-1, \ell = 0) .$$

When we (Richard, Taxil and A.M.) undertook this study we hoped that the same criterion as before, i.e., the sign of $(d/dr)(1/r)(dV/dr)$ would decide whether the levels get either closer and closer with increasing n or more and more spaced. This is not true. We have examples of perturbations of $r^2/4$ such that $(d/dr)(1/r)(dV/dr) > 0$ for which, for some n,

$$2E(n, 0) > E(n+1, 0) + E(n-1, 0) ,$$

while we expected the opposite. In pure power potentials, $V = r^v$, the naïve expectation seems true, i.e.,

$$2E(n,0) \gtrless E(n+1,0) + E(n-1,0) \text{ for } v \gtrless 2 .$$

What we have proved was completely unexpected, although *a posteriori* it was understood. It is as follows:

Theorem:
For

$$V = \frac{r^2}{4} + \lambda v ,$$

λ being sufficiently small, the levels get more and more spaced with increasing n, if

$$\lim_{r \to 0} r^3 v = 0$$

and

$$Z(r) = \frac{d}{dr} r^5 \frac{d}{dr} \frac{1}{r} \frac{dv}{dr} > 0 \ \forall r > 0 . \tag{2.168}$$

If, on the other hand, $Z < 0$, the levels get closer and closer as n increases. However, this is not always true outside the perturbation regime, as we shall see.

Now, why do we need the condition (2.168)? Z is a third-order differential operator in v. There are two obvious vs which make Z equal to zero. These are

$$v = \text{const}, \quad v = r^2 .$$

However, $Z \equiv 0$ admits another solution, i.e.

$$v = \frac{1}{r^2} ,$$

and this is perfectly normal. Adding $v = C/r^2$ to the harmonic oscillator potential is equivalent to shifting the angular momentum by a fixed amount, and since the Regge trajectories of the oscillator are linear and parallel — i.e., $(d/d\ell)(E(n,\ell))$ is constant, independent of n — all levels are shifted by the same amount and equal spacing is preserved. Naturally, we could invent other criteria, since knowledge of all $\ell = 0$ levels does not fix the potential — as all experts on the inverse problem know. There is an infinite-dimensional family of potentials with equal spacing of $\ell = 0$ levels, the parameters being the wave functions at the origin, but we believe that we have here the simplest criterion.

The proof again uses the technique of raising and lowering operators. We want to calculate

$$\lambda\Delta = \delta E_{N+1} + \delta E_{N-1} - 2\delta E_N ,$$

where $E_N = E(N, \ell = 0)$. If $V = r^2/4 + \lambda v$, then

$$\Delta = \int_0^\infty v[u_{N+1}^2 + u_{N-1}^2 - 2u_N^2]dr ; \tag{2.169}$$

u_{N-1} and u_{N+1} are obtained from u_N by using raising or lowering operators:

$$\left(2N + 2 - \frac{r^2}{2} + r\frac{d}{dr}\right)u_N = \sqrt{(2N+2)(2N+3)}u_{N+1}$$

$$\left(2N + 3 - \frac{r^2}{2} - r\frac{d}{dr}\right)u_{N+1} = \sqrt{(2N+2)(2N+3)}u_N . \tag{2.170}$$

Then, one performs a series of integrations by parts with the purpose of transforming Δ into an expression containing explicitly $Z(r)$ defined in (2.168), taking into account the boundary condition $\lim_{r\to 0} vr^3 = 0$. One gets

$$\Delta = \int_0^\infty K(r)\left(\frac{d}{dr}r^5\frac{d}{dr}\frac{1}{r}\frac{dV}{dr}\right)dr , \tag{2.171}$$

where

$$K(r) = \int_r^\infty \frac{dr'}{r'^5}\int_{r'}^\infty r''dr'' \int_{r''}^\infty (u_{N+1}^2 + u_{N-1}^2 - 2u_N^2)dr ;$$

the explicit expression of $K(r)$ is

$$2N(2N+1)(2N+2)(2N+3)K(r) = +u_N'^2\left[\frac{3}{4r^3} + \frac{E}{2r}\right]$$

$$+ u_N^2\left[-\frac{3E}{4r^3} - \frac{3}{16r} + \frac{E^2}{2r} - \frac{Er}{8}\right]$$

$$+ u_Nu_N'\left[-\frac{3}{4r^4} + \frac{E}{2r^2}\right] . \tag{2.172}$$

It is possible (but painful) to prove that (2.172) is everywhere positive. Then Theorem (2.168) follows.

Unfortunately this theorem only holds for perturbations, as several counterexamples show. The best counterexample is obtained by taking one of the partly soluble potentials obtained by Singh, Biswas and Datta [95] and much later by Turbiner [96] of the form $V = r^6 - (2K + 1)r^2$.

For $K = 4$, the energies of the first five levels are given by

$$
\begin{cases}
E_0, \ E_4 = \mp\sqrt{480 + 96\sqrt{11}} \\[2mm]
E_1, E_3 = \mp\sqrt{480 - 96\sqrt{11}} \\[2mm]
E_2 = 0 .
\end{cases}
$$

It is clear that the spacing is not increasing in spite of the fact that Z, calculated from V is positive. However, no counterexample has been found for Z positive and V monotonous increasing.

It may be some time before this conjecture is proved. In the meantime, we shall try to obtain another criterion, based on the WKB approximation. For a monotonous potential, E can be considered to be a continuous function of N, defined by

$$
N - \frac{1}{4} = \frac{1}{\pi} \int_0^\infty \sqrt{(E - V)_+} \, dr . \tag{2.173}
$$

(Here the ground state is given by $N = 1$, not $N = 0$!) Then the spacing is given by dE/dN, and levels get <u>closer</u> with increasing N if

$$
\frac{d}{dN}\left(\frac{dE}{dN}\right) < 0 , \tag{2.174}
$$

i.e., exchanging function and variable:

$$
\frac{d^2N}{dE^2} > 0 . \tag{2.175}
$$

We have, differentiating under the integral sign,

$$
\frac{dN}{dE} = \frac{1}{2\pi} \int_0^{r_t} \frac{dr}{\sqrt{E - V}} , \tag{2.176}
$$

where r_t is the turning point. We cannot differentiate again, because we would get a divergence at r_t. So we assume $V'(r) > 0, \ \forall r \geq 0$, and integrate (2.176) by parts

$$
\frac{dN}{dE} = \frac{1}{2\pi} \int \frac{1}{V'} \frac{V' dr}{\sqrt{E - V}} = \frac{1}{\pi} \frac{\sqrt{E - V(0)}}{V'(0)} - \frac{1}{\pi} \int_0^{r_t} \frac{V''}{V'^2} \sqrt{E - V} \, dr . \tag{2.177}
$$

Hence

$$
\frac{d^2N}{dE^2} = \frac{1}{2\pi} \frac{1}{V'(0)\sqrt{E - V(0)}} - \frac{1}{2\pi} \int_0^{r_t} \frac{V''}{V'^2 \sqrt{E - V}} dr .
$$

A sufficient condition for this to be positive is

$$
V'' < 0 ; \tag{2.178}
$$

however, notice that d^2N/dE^2 does not vanish for $V'' \to 0$, since the first term survives and cannot be zero because $V'(0)$ is not infinite. In fact, for a purely linear potential, it is easy to prove rigorously that the spacing between the energy levels decreases.

We are convinced that Condition (2.178) is sufficient to ensure that the energy level spacing decreases, but it may be some time before a rigorous proof is given.

In QCD, the quark–antiquark potential is believed to be concave [46]. In this way we understand that in the case of the Upsilon ($b\bar{b}$ bound states) and Charmonium ($c\bar{c}$ bound states) we have:

$$M_{\gamma''} - M_{\gamma'} < M_{\gamma'} - M_{\gamma} ,$$
$$M_{\gamma'''} - M_{\gamma''} < M_{\gamma''} - M_{\gamma'} ,$$
$$M_{\psi''} - M_{\psi'} < M_{\psi'} - M_{\psi} .$$

However, the last observed level of the $b\bar{b}$ system, called γ^{IV}, is such that

$$M_{\gamma^{IV}} - M_{\gamma'''} > M_{\gamma'''} - M_{\gamma''} .$$

We therefore have an indication that the naïve one-channel potential model breaks down. However, this level is high above the $B\bar{B}$ threshold

$$M_{\gamma^{IV}} \simeq 10.85 \text{ GeV} \gg 10.55 \text{ GeV}$$

and coupled channel effects or glue contributions or both may be important.

2.4 The wave function at the origin, the kinetic energy, mean square radius etc.

Apart from the energy of the levels, other important quantities are the moments of the wave function, already introduced in Section 2.3 for the nodeless states, the square of the radial wave function at the origin, or, in the case of $\ell \neq 0$, of its ℓ-th derivative, the expectation value of the kinetic energy and the mean square radius of the state. Also of interest are matrix elements between states, such as electric dipole matrix elements etc.

The wave function at the origin plays a role in the decay of positronium or quarkonium. For quarkonium the so-called Van Royen–Weisskopf [97] formula (also proposed by Pietschmann and Thirring, and others) gives

$$\Gamma_{e^+e^-} = 16\pi\alpha e_Q^2 \frac{|\psi(0)|^2}{M^2} , \tag{2.179}$$

where e_Q denotes the charge of the quarks in units of e, and M the total mass of the decaying state. For the decays of the P states of quarkonium, into photons or gluons, analogous formulae exist involving the derivatives of the wave function (see, for instance, Ref. [97]).

The expectation value of the kinetic energy plays a role, for instance, in the mass dependence of a quark–antiquark system for a flavour-independent potential, because of the Feynman–Hellmann theorem:

$$\frac{d}{dm}E(m) = -\frac{\langle T \rangle}{m} , \tag{2.180}$$

where E is the binding energy of the quark–antiquark system and m is the common mass of the two quarks. A trivial change has to be made for unequal masses. The kinetic energy is also related trivially to the total energy for pure power potentials, but there are more general inequalities linking the two for a given sign of the Laplacian if V tends to zero at infinity, and also consequences for the dependence of the spacing of the corresponding levels on the mass.

The importance of the mean square radius is obvious, because it is a measure of the size of the system, and it is desirable to relate it to more accessible quantities like energies.

All these quantities, as we shall see, are linked together by all sorts of inequalities, some of which are rather tight and can be of physical interest.

Returning to the wave function at the origin or its derivative let us remark first that it is, in fact, the limiting case of a moment:

$$\lim_{r \to 0} \frac{(u_{\ell,n}(r))^2}{r^{2\ell+2}} = \lim_{v \to -2\ell-3} \frac{\int (u_{\ell,n}(r))^2 r^v \, dr}{\Gamma(2\ell + 3 + v)} . \tag{2.181}$$

In the special case of $n = 0$, this will allow us to use the arsenal of theorems obtained at the beginning of Section 2.3, but for the time being we shall look specifically at radial excitations, with mostly $\ell = 0$. A first result follows [98]:

Theorem:
If

$$V'' \gtrless 0 \quad \forall r > 0 \quad |u'_{1,0}(0)|^2 \gtrless |u'_{0,0}(0)|^2 . \tag{2.182}$$

Proof:

$u_{1,0}(r)$ has at most one node. By looking at the Wronskian of $u_{0,0}(r)$ and $u_{1,0}(r)$ it is easy to see that $(u_{0,0}(r))^2 - (u_{1,0}(r))^2$ vanishes at least once (because of the normalization) and at most twice. From the Schwinger formula, written for $2m = 1$,

$$(u'(0))^2 = \int_0^\infty u^2(r) \frac{dV}{dr} dr , \tag{2.183}$$

we see that it is impossible to have $(u'_{1,0}(0))^2 = (u'_{0,0}(0))^2$ if V'' is either constantly positive or constantly negative, because then $(u_{1,0}(r))^2 - (u_{0,0}(r))^2$ would vanish only once, at $r = r_0$, and we would have

$$0 = \int_0^\infty \left[(u_{1,0}(r))^2 - (u_{0,0}(r))^2 \right] \left[\frac{d}{dr} V(r) - \frac{dV}{dr}(r_0) \right] dr . \qquad (2.184)$$

If V'' is negative we cannot have $(u'_{1,0}(0))^2 > (u'_{0,0}(0))^2$ because we have the opposite inequality for a Coulomb potential and, in the continuous family of concave potentials,

$$V_\lambda(r) = (1 - \lambda)V(r) - \frac{\lambda}{r} ,$$

one potential would give a zero difference of the wave functions at the origin. For the case $V'' > 0$ a similar argument can be used by taking

$$V_\lambda(r) = (1 - \lambda)V(r) + \lambda r^2 .$$

∎

It is tempting to make the conjecture [99]

$$|u'_{n,0}(0)| \gtrless |u'_{n+1,0}(0)| \quad \text{for} \quad V'' \lessgtr 0 , \quad \forall n , \qquad (2.185)$$

but this conjecture, so far, has resisted all attempts at proof. In the limiting case $V(r) = cr$, we have $|u'_{n,0}(0)| = |u'_{n+1,0}(0)| \; \forall \, n$, according to (2.183). All we can do is to look at the limiting case $V = r + \lambda v$, λ small, V with a given concavity. Then, by a mixture of numerical calculations for small n and use of the asymptotic properties of the Airy functions for large n, we could 'prove' that

$$\lim_{\lambda \to 0} \frac{|u'_{n+1}(0)|^2 - |u'_n(0)|^2}{\lambda} \gtrless 0 \quad \text{if} \quad v'' \gtrless 0 .$$

Nevertheless a weaker, but solid result can be obtained:

Theorem:
If

$$V'' < 0 , \quad V'(r) \to 0, \quad \text{and} \quad V(r) \to \infty,$$
$$\text{for} \; r \to \infty , |u'_{n,0}(0)|^2 \to 0 \; \text{for} \; n \to \infty . \qquad (2.186)$$

Proof:
If $V'' < 0$, V is necessarily monotonous-increasing, because it cannot have a maximum for finite r without going to $-\infty$ for $r \to \infty$. Then the quantity $u'(r)^2 + (E - V)u^2(r)$ is monotonous-decreasing, as one can see by calculating its derivative. Hence

$$|u'(0)|^2 > |u'(r)|^2 \quad \forall r > 0 ,$$

and, by integration,

$$|u(r)|^2 < r^2 |u'(0)|^2 ,$$

but, also,

$$|u(r)|^2 < \frac{|u'(0)|^2}{E - V(r)} .$$

Hence we get, by inserting both inequalities in (2.183):

$$|u'_{n,0}(0)|^2 \left[1 + 2 \int_0^{R_1} rV(r)dr - R_1^2 V(R_1) + \ln \frac{E(n,0) - V(R_2)}{E(n,0) - V(R_1)} \right] < \frac{dV}{dr}(R_2) ,$$
(2.187)

where $0 < R_1 < R_2 < R_T$, and where R_T is the turning point

$$E(n,0) = V(R_T) .$$

We can choose R_1 sufficiently small, but fixed, such that

$$1 + 2 \int_0^{R_1} r|V(r)|dr - R_1^2 V(R_1)$$

is strictly positive. Then we can choose R_2 such that $dV/dr(R_2) < \epsilon$, ϵ arbitrarily small, and finally since $V(r) \to \infty$, $E(n,0) \to \infty$ for $n \to \infty$, and we can make the logarithm in (2.187) arbitrarily small. Therefore, for n big enough, we can make $|u'_{n,0}(0)|^2$ arbitrarily small, under the acceptable extra condition that $\int_0^R r|V(r)|dr$ converges.

In the case of a potential going to zero at infinity but having an infinite number of bound states, we also have the property $|u'_{n,0}(0)|^2 \to 0$ for $n \to \infty$. Let us impose the condition

$$(2 - \epsilon)V(r) + r\frac{dV}{dr} < 0, \quad \text{with} \quad V(r) \to 0 \text{ as } r \to \infty ,$$
(2.188)

which guarantees that the potential decreases more slowly than $r^{-2+\epsilon}$ and has an infinite number of bound states accumulating at zero energy.

We write, using obvious inequalities,

$$|u'_{n,0}(0)|^2 \left[1 - \int_0^R r^2 \frac{dV(r)}{dr} dr \right] < \frac{1}{R} \int_R^{\infty} r\frac{dV}{dr} u_{n,0}^2 dr ,$$

and hence, if R is small enough,

$$|u'_{n,0}(0)|^2 < \frac{\frac{1}{2}\langle T \rangle_n}{R \left[1 - \int_0^R r^2 (dV/dr)dr \right]} ,$$

where $\langle T \rangle_n$ is the average kinetic energy. But, from (2.188)

$$\langle T \rangle_n < \frac{2 - \epsilon}{\epsilon} |E(n,0)| ,$$

and because $E(n,0)$ goes to zero for $n \to \infty$, the desired result follows.

Conversely if $V'' > 0$, $V' > 0$, $V' \to \infty$ for $r \to \infty$, we have $|u'_{n,0}(0)|^2 \to \infty$ for $n \to \infty$. The proof is simple because V cannot be singular at the origin. One gets

$$|u'_{n,0}(0)|^2 \geq \frac{dV}{dr}(R) / \left[1 + \frac{dV(R)}{dr} \int_0^R \frac{dr}{E(n,0) - V(r)} \right] .$$

One can take R sufficiently large to make $dV(R)/dr$ arbitrarily large, and then n big enough to make the denominator arbitrarily close to unity. ∎

A final indication in favour of our conjectures is the WKB approximation. Averaging for large n over the oscillations of the square of the WKB wave function, i.e., replacing

$$u^2 \simeq \frac{1}{(E-V)^{1/2}} \sin^2 \left[\int_0^r \sqrt{E - V(r')} dr' \right] \times \text{const}$$

by

$$\frac{1}{2} \frac{1}{(E-V)^{1/2}} \times \text{const} ,$$

one gets, after some manipulations,

$$\frac{d}{dE} |u'(0)|^{-2} = -\frac{1}{2} \int_0^{R_T} dr \frac{VV''}{E^2 V'^2} \left(1 - \frac{V}{E} \right)^{-1/2} , \tag{2.189}$$

which means that $|u'(0)|$ decreases when E increases for $V'' < 0$, and the reverse for $V'' > 0$. This is a continuous version of our conjecture.

Another approach to the wave function at the origin has been proposed by Glaser [100]. He starts from the matrix element

$$\langle x | e^{-Ht} | y \rangle , \tag{2.190}$$

which for $x = y$ reduces to

$$\langle x | e^{-Ht} | x \rangle = \sum |\psi_n(x)|^2 e^{-E_n t} , \tag{2.191}$$

assuming that the potential is confining and that, hence, we have a purely discrete spectrum. The simplest result is: if the potential is <u>positive</u>,

$$\langle x | e^{-Ht} | x \rangle < \langle x | e^{-H_0 t} | x \rangle , \tag{2.192}$$

which implies, in three dimensions,

$$\sum_{k=1}^N |\psi_k(x)|^2 < \frac{1}{(4\pi)^{3/2}} \frac{e^{E_N t}}{t^{3/2}} ,$$

and, minimizing with respect to t,

$$\sum_{k=1}^{N} |\psi_k(x)|^2 < E_N^{3/2} \left(\frac{e}{6\pi}\right)^{3/2} , \qquad (2.193)$$

which has the same qualitative behaviour as the semiclassical estimate that we shall soon derive.

Now we no longer assume that V is positive, but only that $\int |V_-|^\gamma d^3x$ converges, with $\gamma < 3/2$, V_- being the negative part of the potential. Then we can prove that [100]

$$\left| \frac{\langle x|e^{-Ht}|y\rangle}{\langle x|e^{-H_0t}|y\rangle} - 1 \right| \to 0 \text{ for } t \to 0 ,$$

or

$$\lim_{t\to 0} \left[(4\pi t)^{3/2} \sum_{k=1}^{\infty} |\psi_k(x)|^2 e^{-E_k t} \right] = 1 .$$

Then, thanks to a Tauberian theorem of Karamata, we can prove that

$$\lim_{N\to\infty} \frac{\sum_{k=1}^{N} |\psi_k(x)|^2}{E_N^{3/2}} = \frac{1}{6\pi^2} , \qquad (2.194)$$

which agrees exactly with the semiclassical result obtained by Quigg and Rosner [59] — at least for non-singular central potentials in the WKB approximation:

$$|\psi_N(0)|^2 \sim \frac{1}{4\pi^2} E_N^{1/2} \frac{dE_N}{dN} . \qquad (2.195)$$

Our result (2.194) has the advantage of being valid for any confining potential less singular than $-r^{-2+\epsilon}$ at the origin. For instance, for $V = \ln r$ it gives

$$\sum_{k=1}^{N} |\psi_k(0)|^2 \sim \text{const} \times (\ln N)^{3/2} .$$

In fact, our approach applies to any x, without spherical symmetry, and with mild extra assumptions one can prove

$$\lim_{N\to\infty} \frac{\sum_{k=1}^{N} |\psi_k(x)|^2 - ([E_N - V(x)]^{3/2})/6\pi^2}{E_N^{1/2}} \to 0 , \qquad (2.196)$$

which means that if one fills a confining potential with N independent particles one gets, for large N, the Thomas–Fermi density.

Now we wish to derive inequalities on the expectation values of the kinetic energy as a function of energy differences. Initially this was done by Bertlmann and one of the present authors (A.M.) [101] using sums over intermediate states by a technique somewhat similar to the one used to derive the Thomas–Reiche–Kuhn and Wigner sum rules [97]. But our inequalities hold, in fact, even for non-integer angular momentum, using Regge's analytic continuation. To prove them one must use the Schrödinger differential equation directly. First, we derive an inequality on the mean square radius:

$$E(0, \ell + 1) - E(0, \ell) \leq \frac{2\ell + 3}{\langle r^2 \rangle_{0,\ell}} . \tag{2.197}$$

All we need is an upper bound on $E(0, \ell + 1)$, which can be obtained by taking the variational trial function

$$u_{0,\ell+1} = C r u_{0,\ell} , \tag{2.198}$$

where $u_{0,\ell}$ is the exact wave function of the state $(0, \ell)$. The calculation is straightforward and gives (2.197). It is not a surprise that (2.197) becomes an equality if $V(r)$ is a harmonic oscillator potential, because then the trial wave function given by (2.198) becomes exact.

Now we have a kind of Heisenberg uncertainty inequality:

$$4\langle T \rangle_{0,\ell} \langle r^2 \rangle_{0,\ell} \geq (2\ell + 3)^2 . \tag{2.199}$$

This inequality can by proved by expanding the quadratic form given by

$$\int_0^\infty \left[u'_{0,\ell} - (\ell + 1)\frac{u_{0,\ell}}{r} - \lambda r u_{0,\ell} \right]^2 dr \geq 0 \tag{2.200}$$

and writing that its discriminant is negative. Combining (2.197) and (2.200) we get

$$\langle T \rangle_{0,\ell} \geq \frac{2\ell + 3}{4} [E(0, \ell + 1) - E(0, \ell)] . \tag{2.201}$$

This inequality is completely general and becomes an equality if V is a harmonic oscillator potential. It will be used at the end of this section to calculate the mass difference between the b-quark and the c-quark from experimental data.

If one adds extra conditions one can get inequalities in the opposite direction. For instance, if

$$\frac{d}{dr} \frac{1}{r} \frac{dV}{dr} < 0$$

$$\langle T \rangle_{0,\ell} \leq \frac{2\ell + 3}{4} [E(0, \ell) - E(0, \ell - 1)] . \tag{2.202}$$

The proof of this inequality will be given after we have discussed the 'Common' inequalities [90] between the expectation value of the kinetic energy and that of r^{-2}, which we shall now proceed to present.

Consider first the case where the Laplacian of the potential has a given sign. Then the radial kinetic energy in the state $u = u_{0,\ell}$ can be written as

$$\int_0^\infty u'^2 dr = \int_0^\infty uu' \left(\frac{u'}{u}\right) dr = -\frac{1}{2} \int u^2 \left(\frac{u'}{u}\right)' dr .$$

From Lemma (2.28), we get

$$\int u'^2 dr \gtrless \frac{1}{2} \int \frac{u^2}{r^2}(\ell + 1)dr \tag{2.203}$$

if

$$\Delta V \gtrless 0 .$$

Combining with the centrifugal term we get

$$\langle T \rangle_{0,\ell} \gtrless (\ell + 1) \left(\ell + \frac{1}{2}\right) \langle r^{-2} \rangle_{0,\ell} \tag{2.204}$$

if

$$\Delta V \gtrless 0 \quad \forall r > 0 .$$

A similar treatment can be given for all cases A and B given in Section 2.2 resulting in

$$\langle T \rangle_{0,\ell} \gtrless \left(\ell + \frac{\alpha + 1}{2}\right) (\ell + 1/2) \langle r^{-2} \rangle_{0,\ell} \tag{2.205}$$

if V belongs respectively to sets A, α or B, α.

A special case is $\alpha = 2$, corresponding to a given sign of $(d/dr)(1/r) (dV/dr)$,

$$\langle T \rangle_{0,\ell} \gtrless \left(\ell + \frac{3}{2}\right) \left(\ell + \frac{1}{2}\right) \langle r^{-2} \rangle_{0,\ell} \tag{2.206}$$

if

$$\frac{d}{dr} \frac{1}{r} \frac{dV}{dr} \gtrless 0 \quad \forall r > 0 .$$

Combining this inequality for the case in which

$$\frac{d}{dr} \frac{1}{r} \frac{dV}{dr} < 0$$

with the concavity of $E(0, \ell)$ — see Theorem (2.154) — we get

$$\langle T \rangle_{0,\ell} < \left(\ell + \frac{3}{2}\right) \frac{1}{2} \frac{dE}{d\ell} < \frac{2\ell + 3}{4} [E(0,\ell) - E(0,\ell - 1)] ,$$

which is precisely (2.202).

Let us notice that although we have so far proved these inequalities only for the ground state for a given ℓ, they hold, in fact, <u>in one direction</u>

for arbitrary radial excitations. One has for arbitrary n

$$\langle T \rangle_{n,\ell} \geq (\ell + 1)(\ell + 1/2) \langle r^{-2} \rangle_{n,\ell} \tag{2.207}$$

for $\Delta V \geq 0$, and

$$\langle T \rangle_{n,\ell} \geq (\ell + 3/2)(\ell + 1/2) \langle r^{-2} \rangle_{n,\ell} \tag{2.208}$$

for

$$\frac{d}{dr} \frac{1}{r} \frac{dV}{dr} \geq 0.$$

However, these inequalities are never saturated for $n \geq 1$. Their proof is very simple. Consider a state with n nodes. Consider the interval between two successive nodes $r_k < r < r_{k+1}$, and take as a potential

$$V_{k,k+1} = \left(\frac{r_k}{r}\right)^N - 1 \quad \text{for } r < r_k$$

$$V_{k,k+1} = V \qquad \text{for } r_k < r < r_{k+1} \tag{2.209}$$

$$V_{k,k+1} = \left(\frac{r}{r_{k+1}}\right)^N - 1 \text{ for } r > r_{k+1}.$$

If V satisfies, for instance, $\Delta V > 0$, then $V_{k,k+1}$ also satisfies $\Delta V_{k,k+1} > 0$ for $N > 1$ and hence taking the limit $N \to \infty$ we see that inequality (2.207) holds for the ground state of this limit potential — i.e., for potential V with Dirichlet boundary conditions at r_k and r_{k+1}. Hence

$$\int_{r_k}^{r_{k+1}} \left[u'^2_{n,\ell} + \frac{\ell(\ell + 1)}{r^2} u^2_{n,\ell} \right] dr > (\ell + 1/2)(\ell + 1) \int_{r_k}^{r_{k+1}} r^{-2} u^2_{n,\ell} dr$$

Adding up all such inequalities for all successive intervals gives (2.206). The same method allows one to prove (2.208). Notice that the opposite inequalities <u>are not valid</u> for arbitrary n, because the construction (2.209) is impossible.

Now, since — as we have seen — the wave function at the origin is the limiting case of a moment, we can obtain inequalities connecting the wave function at the origin and the expectation value of the kinetic energy [91, 102].

We have seen in Section (2.3) that

$$\log \frac{\langle r^\nu \rangle_{0,\ell}}{\Gamma((\nu + 2\ell + 3)/\alpha)}$$

is concave (convex) if V belongs to class (A, α) or (B, α), respectively. On the other hand we have

$$\lim_{r \to 0} \left(\frac{u_\ell}{r^{\ell+1}}\right)^2 = \alpha \lim_{\nu \to -2\ell-3} \frac{\langle r^\nu \rangle_{0,\ell}}{\Gamma((\nu + 2\ell + 3)/\alpha)} . \tag{2.210}$$

Applying the convexity considerations to $v = -2\ell - 3$, $v = -2$, and $v = 0$, and using inequality (2.205), one gets

$$\lim_{r \to 0} \left(\frac{u_{0,\ell}}{r^{\ell+1}} \right)^2 \overset{\leq}{>} \alpha \frac{\Gamma\left((2\ell+3)/\alpha \right)^{\ell+1/2}}{\Gamma\left((2\ell+1)/\alpha \right)^{\ell+3/2}} \left(\frac{\langle T \rangle_{0,\ell}}{(\ell+1/2)(\ell+(\alpha+1/2))} \right)^{\ell+3/2}$$

(2.211)

for classes (A, α) and (B, α), respectively, defined by (2.95) and (2.96). Interesting particular cases are, for $\ell = 0$,

$$|u'_{0,0}(0)|^2 \overset{\leq}{>} 4 \langle T \rangle_{0,0}^{3/2}$$

(2.212)

if $\Delta V \gtrless 0$, and

$$|u'_{0,0}(0)|^2 \overset{\leq}{>} \frac{4}{\sqrt{\pi}} \left(\frac{2}{3} \right)^{3/2} \langle T \rangle_{0,0}^{3/2} \simeq 1.2284 \langle T \rangle_{0,0}^{3/2}$$

(2.213)

if

$$\frac{d}{dr} \frac{1}{r} \frac{dV}{dr} \gtrless 0 .$$

An interesting example is $V = r$, corresponding to class A with $\alpha = 3/2$ which gives

$$|u'(0)|^2 < \frac{3}{2} \left(\frac{8}{5} \right)^{3/2} \frac{1}{(\Gamma(2/3))^{3/2}} \langle T \rangle_{0,0}^{3/2} \cong 1.927 \langle T \rangle_{0,0}^{3/2} ,$$

which is not too far from the exact answer $(3/E_0)^{3/2} \cong 1.455$.

These results can be applied to quarkonium physics. The kinetic energy satisfies the Bertlmann–Martin inequality (2.201) which for $n = 0, \ell = 0$ reduces to

$$\langle T \rangle_{0,0} \geq \frac{3}{4} \left[E(n = 0, \ell = 1) - E(n = 0, \ell = 0) \right] .$$

(2.214)

Noticing that the quarkonium potential is <u>concave</u> and <u>increasing</u>, we see that (2.213) applies and in this way we get

$$|\psi(0)|^2 > \frac{1}{\pi^{3/2}} \left[\frac{M_q}{2} [E(\ell = 1) - E(\ell = 0)] \right]^{3/2} .$$

(2.215)

If we substitute this inequality into the Van Royen–Weisskopf formula (2.179) we get, with

$$M(n = 0, \ell = 1) = 3.52 \text{ GeV},$$
$$M(n = 0, \ell = 0) = 3.10 \text{ GeV},$$

$\Gamma_{e^+e^-} > 3.9\text{–}5.2$ keV with a c-quark mass between 1.5 and 1.8 GeV. This is to be compared with

$$\Gamma_{e^+e^-} = 4.7 \pm 0.35 \text{ keV}$$

obtained experimentally. This is almost too good, but we must remember that the Van Royen–Weisskopf formula needs some substantial corrections.

Similarly, for the $b\bar{b}$ system, one gets, with

$$M(n = 0, \ell = 1) = 9.90 \text{ GeV},$$
$$M(n = 0, \ell = 0) = 9.46 \text{ GeV},$$

and a quark mass of 5.174 GeV

$$\Gamma_{e^+e^-} > 0.7 \text{ keV},$$

compared to the experimental value of

$$\Gamma_{e^+e^-} \simeq 1.34 \pm 0.05 \text{ keV}.$$

Now we shall use the previous inequalities on the kinetic energy to study the mass dependence of the binding energy and also the energy level separation.

Expression (2.181) gives the derivative of the binding energy with respect to the common mass of a system made of a particle and an antiparticle bound by a mass-independent potential (in the case of quarkonium a 'flavour-independent' potential). Following Ref. [101] we have

$$E(m) - E(M) = \int_m^M \frac{T(\mu)}{\mu} d\mu. \tag{2.216}$$

Now remember that if we consider the Schrödinger equation

$$\left[-\frac{d^2}{\mu dr^2} + V(r) - E(\mu) \right] u = 0, \tag{2.217}$$

$1/\mu$ enters linearly in the Hamiltonian, and according to a general theorem — see the remark after (2.10) — the ground-state energy of the Hamiltonian is concave with respect to this parameter, i.e.,

$$\frac{d}{d(1/\mu)} \frac{dE}{d(1/\mu)} < 0.$$

Using (2.180) we get

$$\frac{d}{d\mu}(\mu T(\mu)) > 0, \tag{2.218}$$

so that from (2.217) we obtain

$$E(m) - E(M) > mT(m) \left[\frac{1}{m} - \frac{1}{M} \right] \text{ for } M > m.$$

Then adding the quantum numbers which had previously been omitted, and using inequality (2.201), we get

$$E(m, n = 0, \ell = 0) - E(M, n = 0, \ell = 0) >$$

$$\frac{3}{4}\frac{M-m}{M}\left[E(m, n = 0, \ell = 1) - E(m, n = 0, \ell = 0)\right].$$

Now we use

$$M_{Q\bar{Q}}(n, \ell) = 2m_Q + E(m_Q, n, \ell), \qquad (2.219)$$

where m_Q denotes the quark mass and $M_{Q\bar{Q}}$ the total mass of the quark–antiquark system, to get

$$m_b - m_c \geq \frac{M_{b\bar{b}}(n = 0, \ell = 0) - M_{c\bar{c}}(n = 0, \ell = 0)}{2 - \frac{3}{4}(M_{c\bar{c}}(n = 0, \ell = 1) - M_{c\bar{c}}(n = 0, \ell = 0))/m_b}. \qquad (2.220)$$

With $m_b = 5$ GeV this gives us $m_b - m_c > 3.29$ GeV, while the inequality $m_b - m_c > \frac{1}{2}(M_{b\bar{b}} - M_{c\bar{c}})$ gives $m_b - m_c > 3.18$ GeV.

This is as far as we can go without imposing any restriction on the potential. We have seen in Section 2.1 that by scaling for power potentials $V = r^\alpha/\alpha$, one gets from the Schrödinger equation

$$\left(-\frac{1}{\mu}\frac{d^2}{dr^2} + \frac{\ell(\ell+1)}{\mu r^2} + V(r) - E\right)u = 0,$$

$$E(\mu) \sim \mu^{-\frac{\alpha}{\alpha+2}} \quad \text{and} \quad T(\mu) \sim \mu^{-\frac{\alpha}{\alpha+2}}.$$

In this way we see that in the particular cases of a Coulomb potential ($\alpha = -1$), or linear potential ($\alpha = 1$), $T(\mu)$ behaves like μ or like $\mu^{-1/3}$ respectively.

The 'good potentials' for quarkonium are, as we have seen in the introduction, such that

$$\frac{d}{dr}r^2\frac{dV}{dr} > 0 \quad \text{and} \quad \frac{d^2V}{dr^2} < 0. \qquad (2.221)$$

These are particular cases of restrictions of the type

$$\lambda V' + rV'' \gtrless 0 \quad \forall r > 0, \qquad (2.222)$$

the vanishing of (2.222) corresponding to a power potential with $\lambda = 1-\alpha$. From (2.222) we see that

$$(\lambda - 1)\left(V + \frac{1}{2}rV'\right) + \frac{rV'}{2}(3 - \lambda) \qquad (2.223)$$

is increasing or decreasing respectively. Consider now the Schrödinger equations for two masses, m and M, $M > m$. Combining them and integrating we get

$$\frac{u'(M,r)}{u(M,r)} - \frac{u'(m,r)}{u(m,r)} = \frac{\int_0^r [(M-m)V + c]u(M,r')u(m,r')dr'}{u(M,r)u(m,r)} . \tag{2.224}$$

If V is monotonous-increasing, and if $u(M,r)$ and $u(m,r)$ are ground-state wave functions, we see that $u(M,r)/u(m,r)$ is decreasing and takes the value 1 only once at $r = r_0$. If we now take the expectation values of (2.223) for M and m, take the difference and use the virial theorem, we get

$$(\lambda - 1)(E(M) - E(m)) + (3 - \lambda)(T(M) - T(m))$$

$$= \int \left[(\lambda - 1)V + rV' - [(\lambda - 1)V + rV']_{r=r_0} \right] \left[u^2(M,r) - u^2(m,r) \right] dr .$$

Under condition (2.222) we get, respectively,

$$(\lambda - 1)(E(M) - E(m)) + (3 - \lambda)(T(M) - T(m)) \lessgtr 0$$

or, going to the differential form,

$$(1 - \lambda)\frac{T(\mu)}{\mu} + (3 - \lambda)\frac{dT}{d\mu} \lessgtr 0 \tag{2.225}$$

and integrating, we get $T(\mu)\mu^{1-\lambda/3-\lambda}$ decreasing (increasing).

In particular, if the potential has a <u>positive</u> Laplacian and is <u>concave</u>, which is condition (2.221) and which corresponds to $\lambda = 2$ and $\lambda = 0$, we get

$$\begin{cases} T(\mu)/\mu \text{ decreases} \\ \\ T(\mu)\mu^{1/3} \text{ increases .} \end{cases} \tag{2.226}$$

Obviously, (2.226) gives a stronger restriction than (2.220). One obtains

$$E(m, n = 0, \ell = 0) - E(M, n = 0, \ell = 0)$$

$$> \frac{9}{4}\left[1 - \left(\frac{m}{M}\right)^{1/3} \right] [E(m, n = 0, \ell = 1) - E(m, n = 0, \ell = 0)] ;$$

when this is applied to the quarkonium system one gets

$$m_b - m_c > \frac{1}{2}[M_{b\bar{b}}(n = 0, \ell = 0) - M_{c\bar{c}}(n = 0, \ell = 0)]$$

$$+ \frac{9}{8}\left(1 - \left(\frac{m_c}{m_b}\right)^{1/3} \right) [M_{c\bar{c}}(n = 0, \ell = 1) - M_{c\bar{c}}(n = 0, \ell = 0)] .$$

With $m_b = 5$ GeV this gives

$$m_b - m_c > 3.32 \text{ GeV} , \tag{2.227}$$

which is a substantial improvement. If one wants to go further one must abandon rigour, and replace inequality (2.201) for $\ell = 0$ by a realistic guess. In Ref. [101] such a guess was made. It was

$$\langle T \rangle_{0,0} \simeq \tfrac{3}{4}(E(0,1) - E(0,0)) \left[1 + \tfrac{7}{9}c^2\right]$$

$$\text{with}\quad c = [E(1,0) + E(0,0) - 2E(0,1)] / [E(1,0) - E(0,0)]$$

$$\left.\right\} . \quad (2.228)$$

This expression has been manufactured in such a way that the bound (2.201) is always satisfied and is saturated for a harmonic oscillator potential, for which c vanishes. It also fits a pure Coulomb potential. Numerical experiments with the potential $-A/r + Br$ and with pure power potentials give agreement to a few parts per mille.

$T(m_b)$ and $T(m_c)$ can then be estimated:

$$T(m_c) \simeq 0.346 \text{ GeV} ,$$

$$T(m_b) \simeq 0.41 \quad \text{GeV} .$$

Taking the average T in the integration, one gets, for $m_b = 5$ GeV and 5.2 GeV,

$$m_b - m_c \simeq 3.39 \text{ and } 3.38 \text{ GeV} , \tag{2.229}$$

respectively. The sensitivity to m_b is therefore very weak. For another approach involving leptonic width see Ref. [102].

We turn now to another mass dependence, that of the spacing of energy levels [103]. For power potentials this is trivial by scaling [59]. The spacing between two levels with given quantum numbers behaves like $\mu^{-\alpha/(\alpha+2)}$ for $V = r^\alpha/\alpha$. For the separation between the lowest levels, (1,0) and (0,0), (0,1) and (0,0) and, more generally, between two purely angular excitations, we can impose a condition of the type (2.222). It is easy to see that the wave functions $u_{0,L}$ and $u_{0,\ell}$ of two purely angular excitations intersect only once, and that $u_{0,L} < u_{0,\ell}$ for small r. Thus, assuming (2.222) and taking the difference of two expectation values of (2.223) in the states $(0, L)$ and $(0, \ell)$, we get

$$(\lambda - 1)[E(\mu, 0, L) - E(\mu, 0, \ell)] + (3 - \lambda)[T(\mu, 0, L) - T(\mu, 0, \ell)] \gtrless 0$$

if

$$\lambda V' + r V'' \gtrless 0$$

and hence

$$(\lambda - 1)\Delta E - (3 - \lambda)\mu \frac{d\Delta E}{d\mu} \gtrless 0 ,$$

i.e.,

$$\Delta E \times \mu^{\frac{1-\lambda}{3-\lambda}} \text{ decreasing (increasing)} .$$

In particular, if

$$\Delta V > 0 \ (\lambda = 2), \tag{2.230}$$

$$\Delta E \mu \text{ decreases} ;$$

if

$$V'' < 0 \ (\lambda = 0), \tag{2.231}$$

$\Delta E \mu^{1/3}$ increases, as would be expected.

One can also apply these considerations to the 2S–1S separation if the wave functions intersect only once. This is the case for (2.231). But if we only require $\Delta V > 0$ the wave functions may intersect twice. So we obtain the following:

Theorem:
  If $\Delta V > 0$ and $V'' < 0$,
$$\frac{E(\mu, 1, 0) - E(\mu, 0, 0)}{\mu} \text{ decreases and}$$

$$[E(\mu, 1, 0) - E(\mu, 0, 0)] \, \mu^{1/3} \text{ increases with } \mu . \tag{2.232}$$

It will not be a surprise to the reader to learn that these properties are well satisfied by the charmonium and upsilon systems, with

$$E(m_c, 1, 0) - E(m_c, 0, 0) = 3686 - 3097 \text{ MeV} = 589 \text{ MeV}$$

and

$$E(m_b, 1, 0) - E(m_b, 0, 0) = 10023 - 9460 \text{ MeV} = 563 \text{ MeV} .$$

To end this section, we would like to discuss <u>mixed</u> moments — i.e., matrix elements of the type

$$\int_0^\infty u_{n,\ell} r^\nu u_{n',\ell'} dr .$$

First of all, Theorem (2.141) can be adapted trivially. If we call

$$M_{\ell\ell'} = \frac{\int_0^\infty u_{0,\ell} r^\nu u_{0,\ell'} dr}{\Gamma(\ell + \ell' + \nu + 3)},$$

then, if $u_{0,\ell}$ and $u_{0,\ell'}$ are ground-state wave functions with angular momenta ℓ and ℓ' and if the potential has a positive (negative) Laplacian, $\ln M_{\ell\ell'}(\nu)$ is concave (convex) in ν. The analogous theorem, if

$(d/dr)(1/r)(dV/dr)$ has a given sign, also holds: if

$$\frac{d}{dr}\frac{1}{r}\frac{dV}{dr} \gtrless 0$$

$$N_{\ell,\ell'}(v) = \frac{\int_0^\infty u_{0,\ell} r^v u_{0,\ell'} dr}{\Gamma\left((\ell + \ell' + v + 3)/2\right)}$$

is logarithmically concave (convex).

    The theorem can be used to obtain upper and lower bounds on electric dipole matrix elements [104]. Upper bounds are rather trivial:

$$\left[\int u_{0,\ell}(r)u_{0,\ell+1}(r)rdr\right]^2 < \int (u_{0,\ell}(r))^2 r^2 dr ,$$

using Schwarz inequality and the normalization of $u_{0,\ell+1}$. Then, using (2.197), we get

$$\left[\int u_{0,\ell}(r)u_{0,\ell+1}(r)rdr\right]^2 < \frac{E(0,\ell+1) - E(0,\ell)}{2\ell+3} . \tag{2.233}$$

Obtaining a lower bound is trickier. First of all we have the sum rule

$$\int_0^\infty \frac{2\ell+3}{r^2}u_{0,\ell}u_{0,\ell+1}dr = (E(0,\ell+1) - E(0,\ell)) \int_0^\infty u_{0,\ell}u_{0,\ell+1}dr ,$$

obtained by calculating the Wronskian of $u_{0,\ell}$ and $u_{0,\ell+1}$. Then we assume $(d/dr)(1/r)$
$(dV/dr) < 0$, and employ the logarithmic convexity of $N_{\ell,\ell+1}(v)$ for $v = -2$, 0, and 1, making use of the previous identity. Hence we get

$$\left[\int u_{0,\ell}u_{0,\ell+1}rdr\right]^2 > \frac{[\int u_{0,\ell}u_{0,\ell+1}dr]^2 2\Gamma(\ell+5/2)^2/\Gamma(\ell+2)}{E(0,\ell+1) - E(0,\ell)} .$$

It only remains now to get a lower bound on $\int u_{0,\ell}u_{0,\ell+1}dr$, in terms of physically observable quantities. The trick of Common [104] is to take $u_{0,\ell} - \lambda u_{0,\ell+1}$, with $\lambda = [\int u_{0,\ell}u_{0,\ell+1}dr]^{-1}$, as trial function for the first radial excitation. In this way one gets a lower bound on $E(1,\ell)$ which, after some algebraic manipulations gives

$$\left[\int u_{0,\ell}u_{0,\ell+1}dr\right]^2 > \frac{E(1,\ell) - E(0,\ell+1) + 2(\ell+1)\int dr(u_{0,\ell+1})^2/r^2}{E(1,\ell) - E(0,\ell)} .$$

At this point, there are various possibilities:

    (i) keep only the assumption $(d/dr)(1/r)(dV/dr) < 0$, then

$$(2\ell+3)\int \frac{u_{0,\ell+1}^2}{r^2}dr = \frac{dE}{d\ell}(0,\ell+1) < E(0,\ell+2) - E(0,\ell+1) ,$$

which gives

$$\left[\int u_{0,\ell}u_{0,\ell+1}rdr\right]^2 > \frac{2(\Gamma(\ell+5/2))^2}{\Gamma(\ell+2)}$$
$$\times \frac{E(1,\ell) - E(0,\ell+1) + ((2\ell+2)/(2\ell+3))[E(0,\ell+2) - E(0,\ell+1)]}{[E(0,\ell+1) - E(0,\ell)][E(1,\ell) - E(0,\ell)]};$$

(ii) add the assumption $\Delta V > 0$. This will guarantee that the lower bound is non-trivial because $E(1,\ell) > E(0,\ell+1)$. But one can do better by integrating (2.154). One then gets

$$(2\ell+3)\int dr(u_{0,\ell+1})^2/r^2 = \frac{dE}{d\ell}(0,\ell+1) >$$

$$(E(0,\ell+1) - E(0,\ell)) \times \frac{2(\ell+1)^2}{(2\ell+3)(\ell+2)},$$

and hence

$$\left[\int u_{0,\ell}u_{0,\ell+1}rdr\right]^2 > 2\frac{\Gamma(\ell+5/2)}{\Gamma(\ell+2)}$$
$$\times \frac{E(1,\ell) - E(0,\ell+1) + ((\ell+1)^2/((\ell+3)^2(\ell+2)))[E(0,\ell+1) - E(0,\ell)]}{[E(0,\ell+1) - E(0,\ell)][E(1,\ell) - E(0,\ell)]}.$$

For the Upsilon system one finds in this way

$$1.76 \text{ GeV}^{-1/2} < \langle r \rangle_{0,1} < 2.61 \text{ GeV}^{-1/2}.$$

This is not too bad, considering that several of the inequalities used are not optimal.

Up to now, we have considered electric dipole transition moments between states with zero nodes — i.e., with positive radial wave functions. Transition matrix elements between states with $n \neq 0$ in the initial or the final states might vanish. However, there are a few results that one can obtain for matrix elements

$$D_{n,\ell}^{n',\ell'} = \int u_{n,\ell}(r)u_{r',\ell'}(r)rdr, \tag{2.234}$$

where $|\ell - \ell'| = 1$, and $n + n' \leq 2$.

Here we shall not give the details of the proofs which are given in Ref. [105], but only indicate the basic ingredients:

(i) $u_{0,\ell'}(r)/u_{0,\ell}(r)$ is increasing for $\ell' > \ell$;

(ii) $(ru_{0,\ell}/u_{0,\ell+1})$ increases (decreases) if V'/r increases (decreases);

(iii) the sum rule (already used is this section)

$$(E_{n,\ell} - E_{n',\ell'}) \int u_{n,\ell} u_{n',\ell'} dr = \frac{(\ell - \ell')(\ell + \ell' + 1)}{2\mu} \int \frac{1}{r^2} u_{n,\ell} u_{n',\ell'} dr ,$$

where μ is the reduced mass of the particles;

(iv) the sum rule (already quoted in Section 2.1)

$$\mu(E_{n,\ell} - E_{n',\ell'})^2 D_{n,\ell}^{n',\ell'} = \int \frac{dV}{dr} u_{n,\ell} u_{n',\ell'} dr ,$$

if $|\ell - \ell'| \leq 1$;

(v) the fact that if

$$f(v) = \int r^v \rho(r) dr ,$$

the number of zeros of $f(v)$ is less than or equal to the number of zeros of $\rho(r)$.

The results, in a simplified form, are the following:

(1) $D_{0,\ell+1}^{1,\ell}$ cannot vanish, and if we decide that the wave functions are positive near the origin, it is negative.

(2) $D_{0,\ell}^{1,\ell+1} \gtrless 0$ if $(1/r)(dV/dr)$ is increasing (decreasing),
$D_{0,\ell}^{1,\ell+1} = 0$ for an harmonic oscillator potential.

(3) $D_{0,\ell+1}^{2,\ell} \neq 0$ if

$$V = \int_{-3/2}^{-1} \epsilon(\alpha)\rho(\alpha) r^\alpha d\alpha$$

or if

$$V = \int_{1}^{2} \rho(\alpha) r^\alpha d\alpha$$

with

$$\epsilon(\alpha) = \text{sign of } \alpha$$

$$\rho(\alpha) \geq 0 .$$

Conversely there exists a value of α,

$$-1 < \alpha < +1 ,$$

such that

$$D_{0,\ell+1}^{2,\ell} = 0$$

for

$$V = \epsilon(\alpha)r^{\alpha} .$$

(4) $D_{0,\ell}^{2,\ell+1} \neq 0$ for

$$V = \int_{-1}^{+1} \epsilon(\alpha)\rho(\alpha)r^{\alpha}d\alpha .$$

This is perhaps the most interesting non-trivial property since existing potential models for quarkonium belong to this class.

2.5   Relativistic generalizations of results on level ordering

In the previous sections we have obtained a number of results on level ordering of energy eigenvalues belonging to different quantum numbers of n and ℓ, where n denotes the number of nodes of the radial wave function and ℓ the orbital angular momentum, as well as inequalities for pure angular excitations of the Schrödinger equation. A question we are often asked is: What happens if relativistic effects are taken into account? We have already discussed the splitting of multiplets if relativistic corrections are treated in first-order perturbation theory. The next cases to treat are those of the Klein–Gordon and Dirac equations. First of all, it is clear that not all inequalities can be generalized. For some, which become equalities in the case of pure Coulomb potentials in the Schrödinger equation, this is no longer the case if one uses the Klein–Gordon or the Dirac equation. In addition, for the Dirac equation with a vector-like potential, the energy levels depend on the main quantum number N, the total angular momentum J and k, where $|k| = J + (1/2)$, or, equivalently, on ℓ the orbital angular momentum. It has been shown by one of the authors (H.G.) [57] that the degeneracy in k for the Coulomb potential is lifted in such a way that

$$E(N,J,-k) \gtrless E(N,J,k), \qquad \text{if } \Delta V \gtrless 0 \ \forall r > 0, \ k > 0 \qquad (2.235)$$

after perturbation by a potential V.

For example, this means that, in spectroscopic notation, the $Ns_{1/2}$ level is above the $Np_{1/2}$ level if $\Delta V > 0$ for all $r > 0$. However, this result is established only in the first order of perturbation theory and there is no inequality connecting energy levels with the same N but different Js. For the pure Coulomb case we know that for fixed N, the energies increase as J increases.

There is a result, obtained by Palladino and Leal-Ferreira, for the special case of an equal mixture of scalar and vector potentials [106], but this does not tell us what happens in the neighbourhood of a pure Coulomb potential which is vector-like.

We shall first discuss the results for the Klein–Gordon equation

$$[p^2 + (m + S)^2 - (E - V)^2]\psi = 0 \qquad (2.236)$$

and afterwards turn to the Dirac case. For the pure Coulomb case $S = 0$, $V = -(Z\alpha)/r$, with $Z\alpha < 1/2$, we know the solution explicitly, since the term V^2 in (2.236) can be incorporated in the centrifugal term in the reduced radial equation

$$\left(-\frac{d^2}{dr^2} + \frac{\ell'(\ell' + 1)}{r^2} + m^2 - E^2 - 2E\frac{Z\alpha}{r}\right)u = 0, \qquad (2.237)$$

with

$$\left(\ell' + \frac{1}{2}\right)^2 = \left(\ell + \frac{1}{2}\right)^2 - (Z\alpha)^2. \qquad (2.238)$$

Hence, (2.237) looks like a Schrödinger equation with a Coulomb potential, and we get energy levels

$$E(n, \ell) = \frac{m}{\sqrt{1 + \frac{(Z\alpha)^2}{\left[n + \frac{1}{2} + \sqrt{(\ell + \frac{1}{2})^2 - (Z\alpha)^2}\right]^2}}}. \qquad (2.239)$$

Here E denotes the total energy. From (2.239) we see that for fixed $N = n + \ell + 1$, E increases as ℓ increases.

Before stating results on level ordering, we note that the energy levels can still be labelled by n and ℓ and the trivial ordering properties

$$E(n + 1, \ell) > E(n, \ell) \quad \text{and} \quad E(n, L) > E(n, \ell) \text{ for } L > \ell \qquad (2.240)$$

hold. They are indeed satisfied provided that $V \le 0$ and $E(n, \ell) > 0$. Then the coefficient of u, the reduced wave function, in the Klein–Gordon equation is a monotonous function of E, and this suffices to show the first inequality of (2.240). The second follows from the Feynman–Hellmann theorem applied to $dE/d\ell$ and the positivity of $(E - V)$.

In the following we shall take $S = 0$ in Eq. (2.236), however, as will be remarked later, some results also hold for $S > 0$.

Theorem:
If $\Delta V(r) \le 0$, $V(r) < 0$ and $V(r) \to 0$ for $r \to \infty$ and $\lim_{r\to 0} r V(r) > -1/2$, we obtain the strict inequalities

$$E(n, \ell) < E(n - 1, \ell + 1). \qquad (2.241)$$

We show this by regarding the Klein–Gordon equation as a Schrödinger equation dependent on a parameter E:

$$\left[-\frac{d^2}{dr^2} + \frac{\ell(\ell+1)}{r^2} + m^2 - E^2 + (2VE - V^2) \right] u = 0. \qquad (2.242)$$

Under the assumptions made (we restrict ourselves to positive energy states), the Laplacian of the effective potential $2VE - V^2$ is negative because

$$\Delta V^2 = 2 \left(V\Delta V + \left(\frac{dV}{dr} \right)^2 \right). \qquad (2.243)$$

Now, since $\Delta V < 0$ and $V < 0$, rV has a negative limit for $r \to 0$, meaning that the effective potential has a $-1/r^2$ singularity. Our general result of Ref. [56] can be extended to potentials such that $\lim_{r\to 0} r^2 V(r) > -1/4$ and can therefore be applied. On the other hand, the non-linear dependence of (2.242) on the energy can be overcome by using a fixed point argument.

The weakness of this result is that there is no counterpart for $\Delta V \geq 0$. To circumvent this difficulty we follow a procedure analogous to that used in (2.237)–(2.239), i.e., we transfer a part of the potential into the angular momentum term. We define, in analogy to the Coulomb case,

$$- Z\alpha = \lim_{r\to\infty} r\, V(r) \qquad (2.244)$$

and keep the definition of ℓ' (2.238). Then the Klein–Gordon operator becomes

$$\left(-\frac{d^2}{dr^2} + \frac{\ell'(\ell'+1)}{r^2} + m^2 - E^2 + \tilde{V} \right) \qquad (2.245)$$

with

$$\tilde{V} = 2EV - V^2 + \frac{(Z\alpha)^2}{r^2}.$$

Next it can be shown that:

Lemma:
If $\Delta V \gtrless 0 \; \forall r > 0$, $V(r) < 0$, $\lim_{r\to\infty} r\, V(r) = -Z\alpha$, then

$$\Delta \tilde{V} \gtrless 0 \qquad \text{and} \qquad V^2 \gtrless (Z\alpha)^2/r^2. \qquad (2.246)$$

After the change of variables $\tilde{E}(n, \ell') = E(n, \ell)$ we may formulate the following two-sided theorem:

Theorem:
If $\Delta V \gtrless 0$, $V < 0$, $\lim_{r\to\infty} r\, V(r) = -Z\alpha > -\frac{1}{2} - \ell$,

$$\tilde{E}(n, \ell') \gtrless \tilde{E}(n - k, \ell' + k). \qquad (2.247)$$

Remark:

When $\Delta V \leq 0$, a prerequisite of this theorem is that the Klein–Gordon equation make sense — i.e., that $\lim_{r\to 0} r\, V(r) > -\frac{1}{2} - \ell$.

It is straightforward to translate this result into the variable ℓ

$$E(n,\ell) \gtrless E(n-k, \ell + \Delta(\ell,k))$$

with

$$\sqrt{(\ell + \frac{1}{2})^2 - (Z\alpha)^2} + k = \sqrt{(\ell + \Delta + \frac{1}{2})^2 - (Z\alpha)^2}\,. \tag{2.248}$$

In general, Δ will not be an integer if k is one. We have $0 < \Delta(\ell,k) < k$ and in that sense (2.241) is a consequence of (2.247). Indeed, for $\Delta V < 0$, etc., we have

$$E(n,\ell) < E(n-k, \ell + \Delta(\ell,k)) < E(n-k, \ell+k)\,, \tag{2.249}$$

because of the monotonicity of the energies with respect to the angular momentum, which still holds for the Klein–Gordon equation for negative potentials since

$$\frac{dE(n,\ell)}{d\ell}\,\langle E - V \rangle = \left(\ell + \frac{1}{2}\right)\left\langle \frac{1}{r^2} \right\rangle. \tag{2.250}$$

We note that in many situations $\Delta(\ell,k)$ is very close to k.

If $Z\alpha < 1/2$ (which is required for $\ell = 0$) we have $k - 1/2 < \Delta(\ell,k) < k$; if $Z\alpha < 1$, $\ell \geq 1$, we have $k - 0.4 < \Delta(\ell,k) < k$.

Using these remarks we can transform (2.247) into a result connecting only physical angular momenta. If $\Delta V > 0$, $V < 0$, $\lim_{r\to\infty} V(r)r > -1/2$,

$$E(n,\ell) > E(n-k, \ell+k+1)\,. \tag{2.251}$$

This is somewhat crude and does not reduce to the correct limit if V becomes a pure Coulomb potential. However, we can do better by proving a relativistic analogue of the result obtained by Martin and Stubbe [50]:

Theorem:

If $\Delta V \leq 0$, $V(r) < 0$, $\lim_{r\to\infty} r\, V(r) = -Z\alpha > -\frac{1}{2} - \ell_1$,

$$\left(\frac{m^2}{\widetilde{E}^2(0,\ell_1')} - 1\right)(\ell_1' + 1)^2 > \left(\frac{m^2}{\widetilde{E}^2(0,\ell_2')} - 1\right)(\ell_2' + 1)^2 \quad \text{for} \quad \ell_2' > \ell_1'\,.$$

If $\Delta V \geq 0$, $V(r) < 0$, $\lim_{r\to\infty} r\, V(r) = -\ell_1 - \frac{1}{2}$,

$$\left(\frac{m^2}{\widetilde{E}^2(n,\ell_1')} - 1\right)(\ell_1' + 1)^2 > \left(\frac{m^2}{\widetilde{E}^2(n,\ell_2')} - 1\right)(\ell_2' + 1)^2 \quad \text{if} \quad \ell_2' > \ell_1'\,.$$

$$\tag{2.252}$$

The proofs involve a chain of inequalities: the virial theorem for the initial Klein–Gordon equation and a modified one combined with the fact that $V < 0$ and ΔV has a given sign:

$$E(E - \langle V \rangle) \gtrless m^2 \quad \text{for } \Delta V \lessgtr 0 \,,$$

$$\langle \tilde{T} \rangle_{n,\ell'} \gtrless -E \langle V \rangle \quad \text{for } \Delta V \lessgtr 0 \,, \tag{2.253}$$

where $\langle \tilde{T} \rangle_{n,\ell'}$ is the expectation value of $-(d^2/dr^2) + (\ell'(\ell' + 1)/r^2)$. In addition, the Feynman–Hellmann relation for $\tilde{E}$ as a function of ℓ',

$$2 \langle \tilde{E} - V \rangle \frac{d\tilde{E}(n,\ell')}{d\ell'} = (2\ell' + 1) \left\langle \frac{1}{r^2} \right\rangle \,, \tag{2.254}$$

and an inequality connecting the total kinetic energy and the angular kinetic energy established in the Schrödinger case by Common

$$\langle \tilde{T} \rangle_{n,\ell'} > \left(\ell' + \frac{1}{2} \right) (\ell' + 1) \left\langle \frac{1}{r^2} \right\rangle_{n,\ell'} \quad \text{for } \Delta V > 0, \text{ etc.} \,,$$

$$\langle \tilde{T} \rangle_{0,\ell'} < \left(\ell' + \frac{1}{2} \right) (\ell' + 1) \left\langle \frac{1}{r^2} \right\rangle_{0,\ell'} \quad \text{for } \Delta V < 0, \text{ etc.} \tag{2.255}$$

have been used. The last inequalities imply an asymmetry since the last one holds only for $n = 0$ if $\Delta V < 0$. The first one holds for any n, but is optimal in the Coulomb case only for $n = 0$.

Now we can combine (2.247) and (2.252) with $\ell_1 = \ell + \Delta(\ell, k), \ell_2 = \ell + k$ and obtain inequalities, involving only physical angular momenta which reduce to the normal one in the non-relativistic limit, which give for the case $k = 1$, for a potential with a positive Laplacian,

$$\left(\frac{m^2}{E^2(n,\ell)} - 1 \right) < \left(\frac{m^2}{E^2(n-1, \ell+1)} - 1 \right) \left(\frac{\frac{1}{2} + \sqrt{(\ell + \frac{3}{2})^2 - (Z\alpha)^2}}{\frac{3}{2} + \sqrt{(\ell + \frac{1}{2})^2 - (Z\alpha)^2}} \right)^2 \,. \tag{2.256}$$

For a negative Laplacian we have the opposite inequality, but it holds only for $n = 1$. And then, in any case, we have (2.241).

One may wonder about the field of application of these results. It is tempting to apply them to pionic atoms. Unfortunately in this case, however, and contrary to the case of muonic atoms, the major distortion is not due to the size of the nucleus, but to strong interactions, including absorption by the nucleus. Quarkonium has also been described by Kang and Schnitzer [107] with the help of the Klein–Gordon equation, but a pure vector confining interaction led to metastable energy levels.

Probably the most interesting conclusion is that we have found a case in which the obstacles in going from a non-relativistic situation to a relativistic situation are not insurmountable.

We next turn to the Dirac case in which we have results, which are valid in first-order perturbation theory about a vector-like Coulomb interaction [57] and a partial non-perturbative result about the order of levels within a multiplet [108].

For the Dirac operator the Coulomb potential again plays a special role. Although the spin-orbit coupling leads to the fine structure splitting, a certain two-fold degeneracy between levels remains. Our result concerns the splitting of this degeneracy.

The standard treatment of the Dirac–Coulomb problem is rather involved. Therefore a treatment using the symmetry responsible for the degeneracy is more appropriate. We note that in our previous proofs concerning non-relativistic spectra we used supersymmetry in an essential way. In addition, the ladder operators entering the factorization of the Schrödinger–Coulomb Hamiltonian are related to the Runge–Lenz vector. There exists a generalization of this vector for the relativistic situation. In addition, there exists a supersymmetric factorization which is very suitable for studying the level splitting.

We now start to review briefly the algebraic factorization method which allows us to obtain the spectrum and eigenfunctions of the relativistic Coulomb Hamiltonian

$$H = \vec{\alpha} \cdot \vec{p} + \beta m - \frac{\gamma}{r}, \qquad \vec{\alpha} = \begin{pmatrix} 0 & \vec{\sigma} \\ \vec{\sigma} & 0 \end{pmatrix}, \qquad \beta = \begin{pmatrix} 1 & 0 \\ 0 & -1 \end{pmatrix}. \qquad (2.257)$$

Here c is taken as unity $\gamma = Ze^2$, $\vec{\sigma}$ are the Pauli matrices and $\mathbf{1}$ is the 2×2 unit matrix. H acts on four-component spinors with wave functions which are square integrable over $\mathbf{R}^3$. We need the normalized wave functions and a mapping between eigenfunctions belonging to degenerate levels.

Rewriting (2.257) in radial coordinates gives a 2×2 matrix operator

$$h_k = \begin{pmatrix} -\dfrac{\gamma}{r} + m & -\dfrac{d}{dr} + \dfrac{k}{r} \\ \dfrac{d}{dr} + \dfrac{k}{r} & \dfrac{\gamma}{r} - m \end{pmatrix}, \qquad h_k \begin{pmatrix} G_k \\ F_k \end{pmatrix} = E_k \begin{pmatrix} G_k \\ F_k \end{pmatrix}, \qquad (2.258)$$

which acts on two-component wave functions; $k = \pm 1, \pm 2, \ldots$ where $|k| = J + 1/2$ denotes the eigenvalue of the operator $-\vec{\sigma} \cdot \vec{L} - 1$, which commutes with H. There exists a simple transformation [109] from (G_k, F_k) to $(\widetilde{G}_k, \widetilde{F}_k)$,

$$\begin{pmatrix} \widetilde{G}_k \\ \widetilde{F}_k \end{pmatrix} = \begin{pmatrix} k + s_k & -\gamma \\ -\gamma & k + s_k \end{pmatrix} \begin{pmatrix} G_k \\ F_k \end{pmatrix}, \qquad s_k^2 = k^2 - \gamma^2, \qquad (2.259)$$

which allows (2.258) to be rewritten in a supersymmetric factorized way.

We introduce first-order differential operators

$$A_0^\pm = \pm\frac{d}{d\rho} + \frac{s_k}{\rho} - \frac{\gamma}{s_k}, \qquad \rho = E_k r, \tag{2.260}$$

where the index k has been suppressed. $\tilde{G}_k$ and $\tilde{F}_k$ fulfil the relations

$$A_0^+ \tilde{G}_k = \left(\frac{k}{s_k} + \frac{m}{E_k}\right)\tilde{F}_k, \qquad A_0^- \tilde{F}_k = \left(\frac{k}{s_k} - \frac{m}{E_k}\right)\tilde{G}_k. \tag{2.261}$$

The discrete spectrum of H consists of a unique ground state for $k = -1, -2, -3, \ldots$ and a sequence of two-fold degenerate levels for $k = \pm 1, \pm 2, \ldots$. For negative k the ground states are obtained from the conditions $A_0^- \tilde{F}_k = 0$, $\tilde{G}_k = 0$. This leads to

$$\tilde{F}_k = \rho^{s_k} e^{-\gamma\rho/s_k}, \qquad E(0,k) = \frac{m}{\sqrt{1 + \dfrac{\gamma^2}{k^2 - \gamma^2}}}, \qquad k = -1, -2, \ldots. \tag{2.262}$$

No normalizable solutions for $A_0^- A_0^+$ exist at this energy. However, all the other eigenstates of $A_0^- A_0^+$ and $A_0^+ A_0^-$ occur in pairs. They can be obtained by the factorization method [109], since there are identities of the form

$$A_n^- A_n^+ - \frac{\gamma^2}{(s_k + n)^2} = A_{n+1}^+ A_{n-1}^- - \frac{\gamma^2}{(s_k + n + 1)^2},$$

$$A_n^\pm = \pm\frac{d}{d\rho} + \frac{s_k + n}{\rho} - \frac{\gamma}{s_k + n} \qquad \forall n \in \mathbf{N}. \tag{2.263}$$

The n-th eigenfunctions can be obtained from the solutions of

$$A_n^- \chi_n = 0, \qquad \chi_n = \rho^{n+s_k} e^{-\gamma\rho/(s_k + n)} \tag{2.264}$$

by forming $\tilde{F}_k = A_0^+ \ldots A_{n-1}^+ \chi_n$ and evaluating $\tilde{G}_k$ from Eq. (2.261). G_k and F_k are then obtained by inverting (2.259). They correspond to eigenvalues

$$E(n,k) = \frac{m}{\sqrt{1 + \dfrac{\gamma^2}{(s_k + n)^2}}}$$

$$\begin{aligned} n &= 0, &k &= -1, -2, \ldots \\ n &= 1, 2, \ldots, &k &= \pm 1, \pm 2, \ldots \end{aligned} \qquad s_k = \sqrt{k^2 - \gamma^2}, \tag{2.265}$$

which are doubly degenerate for $n \geq 1$.

In the following we fix $n \geq 1$ and $|k|$ and denote $G_{\pm|k|}$ and $F_{\pm|k|}$ by $G_\pm$ and $F_\pm$, respectively. Similarly, we denote $\tilde{G}_{\pm|k|}$ and $\tilde{F}_{\pm|k|}$ by $\tilde{G}_\pm$ and $\tilde{F}$. Note that $\tilde{F}_{\pm|k|}$ does not depend on the sign of k. Our result may be quoted as follows:

Theorem:

The Hamiltonian $H = \vec{\alpha} \cdot \vec{p} + \beta m - \gamma/r + \lambda V(r)$ has a discrete spectrum $E(n, k, \lambda)$, labelled by $k = \pm 1, \pm 2, \ldots$, which is the eigenvalue of $-(\vec{\sigma} \cdot \vec{L} + 1)$ and is related to the total spin quantum number J by $|k| = J + 1/2$ and labelled by $n = 0, 1, 2, \ldots$, which counts states with fixed k and is related to the number of nodes of the radial wave functions. For $n \geq 1$, we have $E(n, k, 0) - E(n, -k, 0) = 0$ and

$$\lim_{\lambda \to 0} \frac{E(n, -k, \lambda) - E(n, k, \lambda)}{\lambda} \gtrless 0 \quad \text{if } \Delta(r) \gtrless 0 \,\forall r \neq 0. \tag{2.266}$$

To prove this we have to determine the sign of

$$\Delta E = \frac{\langle G_+|VG_+\rangle + \langle F_+|VF_+\rangle}{\langle G_+|G_+\rangle + \langle F_+|F_+\rangle} - \frac{\langle G_-|VG_-\rangle + \langle F_-|VF_-\rangle}{\langle G_-|G_-\rangle + \langle F_-|F_-\rangle} \tag{2.267}$$

where $\langle .|. \rangle$ denotes the scalar product in $L^2(\mathbf{R}^+, dr)$.

Two steps remain. Firstly, we have to evaluate the normalization integrals entering (2.267), and secondly, we have to find a mapping relating (G_+, F_+) to (G_-, F_-). With the help of a recursion relation the first task is solved with [57]

$$\frac{\langle F_\pm|F_\pm\rangle + \langle G_\pm|G_\pm\rangle}{\langle \widetilde{F}|\widetilde{F}\rangle} = \frac{m^2}{s_k E_k^2} \frac{1}{(k \pm s_u)(k/s_k \mp m/E_k)}. \tag{2.268}$$

The second, more important, step consists in obtaining ladder operators mapping from states with one sign of k to states with the opposite sign. By explicit calculation we find that the operator Λ_k intertwines between h_k and h_{-k}, where

$$\Lambda_k = \begin{pmatrix} -\dfrac{d}{dr} - \dfrac{k}{r} + \dfrac{m\gamma}{k} & \dfrac{\gamma}{r} \\[2ex] -\dfrac{\gamma}{r} & -\dfrac{d}{dr} + \dfrac{k}{r} + \dfrac{m\gamma}{k} \end{pmatrix} \tag{2.269}$$

and $h_{-k}\Lambda_k = \Lambda_k h_k$. From relation (2.258) we obtain

$$\begin{pmatrix} G_- \\ F_- \end{pmatrix} = \alpha_k \Lambda_k \begin{pmatrix} G_+ \\ F_+ \end{pmatrix}, \qquad \alpha_k = \sqrt{\frac{k + s_k}{k - s_k}} \frac{k}{s_k E_k(k/s_k + m/E_k)}. \tag{2.270}$$

This mapping allows simplification of the expression for ΔE.

Taking a few simple steps like partial integration and using the fact that for a pure Coulomb-like perturbation ΔE has to vanish, we arrive at an expression which has a definite sign if the Laplacian of the potential has a definite sign.

Remarks:

If we assume the validity of the Hartree approximation in the relativistic framework, we can expect from this result the splittings

$$2P_{1/2} > 2S_{1/2}, \quad 3P_{1/2} > 3S_{1/2}, \quad 3D_{3/2} > 3P_{3/2}, \quad \text{etc.} \tag{2.271}$$

for alkaline atoms and their isoelectronic sequences. This is because the outer electron should move in a potential created by a point-like nucleus and the electron cloud of the closed inner shells, which has a negative Laplacian.

The reverse situation occurs in muonic atoms, where at the most naïve level the muon is submitted to the extended charge distribution of the nucleus, thought to be made of indissociable protons and neutrons. Hence by Gauss's law the Laplacian of the potential is positive. Also, the Lamb shift may be viewed as coming from an effective potential with a positive Laplacian, giving the splitting $2P_{1/2} < 2S_{1/2}$. Unlike the Schrödinger case this result still awaits a generalization beyond perturbation theory.

Our next result is that in the atomic case: if $\Delta V < 0$, and if the potential is purely attractive — i.e., $-(\gamma/r) + \lambda V(r) \leq 0$ — one has the complete ordering of levels for fixed principal quantum number N, at least for perturbations around the Coulomb potential. That is, in spectroscopic notation, we have

$$2P_{3/2} > 2P_{1/2} > 2S_{1/2}$$

$$\tag{2.272}$$

$$3D_{5/2} > 3D_{3/2} > 3P_{3/2} > 3P_{1/2} > 3S_{1/2}$$

etc.

The new results are the first inequality in the first line, and the first and the third of the second chain of inequalities, which, as we shall show, are generally true if the potential $-(\gamma/r) + \lambda V(r)$ is purely attractive and monotonously increasing (in our case the latter condition is guaranteed by the hypotheses $\Delta V < 0$). The other inequalities are taken from Ref. [57]. We shall this time label the eigenvalues by the principal quantum number $N = 1, 2, ...$, the total angular momentum $J = 1/2$, $3/2$, ... and the quantum number $k = \pm 1, \pm 2, ...$ In the general case the inequality we shall prove is the following:

$$E_{N,J=L-1/2,|k|} < E_{N,J=L+1/2,-|k|-1}, \tag{2.273}$$

which means that for fixed N and L the energy increases as the total angular momentum J increases.

It was shown in Ref. [50] that in the expansion of Dirac eigenvalues to order $1/c^2$ the inequality (2.273) is always satisfied if $\Delta V \leq 0$, even if the potential is not purely attractive, since the splitting equals the following expectation value for Schrödinger wave functions:

$$\frac{2L+1}{4m^2}\left\langle\frac{\gamma}{r^3}+\frac{1}{r}\frac{d}{dr}V(r)\right\rangle_{N,L}. \tag{2.274}$$

This expression can be shown to have a definite sign, if $\Delta V \leq 0$, by using the sum rule

$$\left\langle\frac{\gamma}{r^2}+\frac{dV(r)}{dr}-\frac{2L(L+1)}{r^3}\right\rangle_{N,L}=0. \tag{2.275}$$

The eigenvalue problem for the Hamiltonian obtained from (2.256) by adding a general potential V to the Coulomb part may be reduced by a standard procedure [110] to a system of two coupled ordinary differential equations,

$$\left(-\frac{\gamma}{r}+\lambda V+m\right)G-F'+\frac{k}{r}F=E_{N,J,k}G\,,$$

$$\left(-\frac{\gamma}{r}+\lambda V-m\right)F+G'+\frac{k}{r}G=E_{N,J,k}F\,, \tag{2.276}$$

where $'$ denotes d/dr and $E_{N,J,K}$ denotes the positive eigenvalue.

Theorem:
We impose the following conditions on the potential $W \equiv -(\gamma/r)+\lambda V$ in Eq. (2.276):

C1: W is purely attractive, i.e., $W \leq 0$;

C2: W is strictly monotonous in r, i.e., $W' > 0$;

C3: $\lim_{r\to 0} rV(r) = 0$ and $\gamma < 1$.

Then $E_{N,J,|k|} < E_{N,J+1,-|k|-1}$.
The last condition means that (without loss of generality) the Coulomb singularity appears only in $-\gamma/r, \gamma < 1$, and guarantees the existence of Coulomb eigenvalues. Since $W \leq 0$, we may obtain a single second-order differential equation for G:

$$-G''-\frac{W'}{E_{N,J,k}+m-W}G'+\frac{k(k+1)}{r^2}G-(E_{N,J,k}-W)^2G+m^2G$$

$$-\frac{k}{r}\frac{W'}{E_{N,J,k}+m-W}G=0\tag{2.277}$$

Doing this for any pair $k = |k| > 0$ and $-|k| - 1$ we observe that the orbital-angular-momentum-like term is the same.
First of all we show that the monotonicity of W implies that

$$E_{N,J,|k|} \neq E_{N,J+1,-|k|-1}\,. \tag{2.278}$$

In fact, suppose that

$$E_{N,J,|k|} = E_{N,J+1,-|k|-1} \equiv E \ .$$

The idea is to construct a suitable eigenvalue problem depending on a certain parameter η which 'interpolates' between these two states. The basic trick is that, in view of the assumption that the energies are the same, we may take $E_{N,J,k}$ as a constant in certain places in Eq. (2.277). In addition, the coefficient of the $1/r^2$ term, $k(k+1)$, remains invariant under the transformation $|k| \to -|k| - 1$. We shall replace Eq. (2.277) by a linear eigenvalue equation which depends on a continuous parameter η. If we label the corresponding eigenfunctions $G_\eta, \eta = |k|$ and $\eta = -|k| - 1$, they satisfy an eigenvalue problem of the form

$$\tilde{H} G_\eta - \frac{\eta}{r} \frac{W'}{E + m - W} G_\eta = \epsilon_\eta G_\eta \ , \tag{2.279}$$

where

$$\tilde{H} = -\frac{d^2}{dr^2} - \frac{W'}{E + m - W} \frac{d}{dr} + \frac{k(k+1)}{r^2} - W^2 + 2EW + m^2 \tag{2.280}$$

and

$$\epsilon_\eta = E^2 \quad \text{for} \quad \eta = |k| \quad \text{and} \quad \eta = -|k| - 1 \ .$$

It should be stressed that, in view of condition C3, the eigenvalue problem (2.279) is well posed and $\tilde{H}$ is a well-defined operator. In particular, there is no $1/r^2$ singularity with negative sign in the effective potential at the origin.

Now we consider η as a parameter varying between $-|k| - 1$ and $|k|$. Since $G_{|k|}$ and $G_{-|k|-1}$ have the same number of nodes (see, for example, Ref. [111]) we may find a curve $\eta \to (G_\eta, \epsilon_\eta)$ varying between $(G_{|k|}, E^2)$ and $(G_{-|k|-1}, E^2)$. On the other hand, the Feynman–Hellmann relation yields

$$\frac{d\epsilon_\eta}{d\eta} = \int_0^\infty \frac{1}{r} \frac{W'}{(E + m - W)^2} G_\eta^2 dr \bigg/ \int_0^\infty \frac{G_\eta^2}{E + m - W} dr > 0 \ , \tag{2.281}$$

which gives the desired contradiction.

Now suppose $E_{N,J,|k|} > E_{N,J,+1,-|k|-1}$. In this case, we study the Dirac equation with a potential W_a, which is a linear combination of the form

$$W_a = aW_c + (1 - a)W \quad a\epsilon[0, 1] \ ,$$

where W_c is another Coulomb potential such that W_a satisfies conditions C1–C3. For the pure Coulomb case ($a = 1$), however, it is known that $E_{N,J,|k|} < E_{N,-J+1,-|k|+1}$. Since W_a is continuous in a, there is a critical value a_0 such that $E_{N,|k|} = E_{N,-|k|-1}$ and the argument presented above will apply.

Remarks:

The proof of the inequality $E_{N,J,|k|} < E_{N,J,+1,-|k|-1}$ is, in fact, non-perturbative and uses only the fact that W is monotonous-increasing (with some smoothness assumptions). We see that the predictions about the splittings in the $1/c^2$ expansion also hold for the 'full' relativistic equation.

As a consequence, taking into account the perturbative result of Ref. [57], we have the complete ordering of levels for the Dirac equation if $\Delta V < 0$. Notice also that assumptions C1 and C2 follow from the combination of $\Delta V < 0$ and $\lim_{r \to \infty} |rV(r)| < \infty$. Specifically, within a Coulomb multiplet, the energies increase for fixed L and increasing J, and for fixed J and increasing L. Overall, the trend is the same as for the case of the Klein–Gordon equation, where the energies increase with increasing L [56]. The kind of ordering we have obtained can be observed in the spectra of a 'one-electron system', such as the LiI isoelectronic sequence [53, 112].

2.6   The inverse problem for confining potentials

Reviewing previous results

After having established a number of relations between the relevant quantities observed in quarkonium systems, one can idealize further and study the inverse problem. The inverse problem for potentials going to zero at infinity in the one-dimensional and radial symmetric case is well worked out and a large amount of literature about the subject is available: for a recent study of this see Chadan and Sabatier [113]. Here we are considering a situation which has not been treated very extensively in the literature. We assume that not only a finite number of levels and slopes for the reduced wave function at the origin are given, but also an infinite sequence of them. Is the potential then determined uniquely? Is there a way to reconstruct it from the data?

Such a programme has been undertaken in a practical way by Quigg, Rosner and Thacker (QRT) [114]. Their work may be summarized as follows. First they consider one-dimensional confining potentials, symmetric with respect to the origin, and solve an approximate problem: they give themselves the first n levels, choose a zero of energy somewhere between the n-th and the $(n+1)$-th levels, and solve the inverse problem as if the potential were vanishing at infinity. The reflection coefficient at positive energies, which is needed to solve the classical one-dimensional problem, is set equal to zero; the potential can then be obtained explicitly originating from a pure superposition of solitons. It is hoped by increasing n to get a sequence of potentials closer and closer to the true one; QRT have made impressive numerical experiments for typical potentials and found a rapid convergence with n. In the second part of their work they deal with radial potentials $V(r)$ and S waves, noticing that when one uses the reduced wave function $u = r\psi\sqrt{4\pi}$, the energy levels coincide with the odd-parity levels inside a symmetric one-dimensional potential $V(|x|)$. The even levels are missing, but on the other hand the $u_i'(0)$ are given; for small i, the algebraic equations relating the missing energies and $u_i'(0)$ can be solved numerically.

In principle, the following problems remain:

(i) Does knowledge of the infinite sequence of $\ell = 0$ energy levels E_i and $|u_i'(0)|^2$ uniquely fix the potential?

(ii) Can one prove the convergence of the QRT procedure?

(iii) Can one obtain the even-parity energy levels from knowledge of E_i and $u_i'(0)$ associated with the odd levels, which are the physical ones?

We have answered questions (i) and (iii) in Ref. [115]. Question (ii) has been dealt with by the Fermilab group [116, 117]. Although there is no general convergence proof, the practical procedure shows a certain kind of approach to the given potential.

We have found two possibilities to solve problems (i), and (iii): the first was to use as much as possible the previous work summarized by Gasymov and Levitan [118]. Here we should mention that in one of the first works on potentials with equivalent spectra [119], the discrete case had already been discussed. There the explicit construction of such potentials is given.

In Ref. [118] the uniqueness question for confining potentials with the 'wrong' boundary conditions, $u'(r = 0) = hu(r = 0)$ with h finite, is discussed, whereas we are interested in $u(0) = 0$. However, we shall show how it is possible to relate our problem to the other one.

We have tried to avoid the use, specified in the lemmas and theorems below, of the older work and succeeded at the price of the 'technical' assumption that acceptable potentials should fulfil.

Finally, we shall make a few comments about the inverse problem when all the ground-state energies are given as a function of the angular momentum.

From the Gasymov–Levitan review we learn:

Theorem:
Let $u(r, E)$ be a solution of the Schrödinger equation with boundary

conditions

$$- u'' + (V - E)u = 0, \quad u'(0, E) = 0, \quad u(0, E) = 1 . \tag{2.282}$$

Now given normalization constants γ_i' and energies E_i' with

$$(\gamma_i')^{-1} = \int_0^\infty dr u^2(r, E_i') , \tag{2.283}$$

one may define a spectral density $\sigma(E)$,

$$\sigma(E) = \begin{cases} \rho(E) & \text{for } E < 0 \\ \rho(E) - \frac{2}{\pi}\sqrt{E} & \text{for } E > 0 \end{cases} \tag{2.284}$$

$$d\rho(E) = dE \sum \gamma_i' \delta(E - E_i') ,$$

and one is able to reconstruct the potential by solving the Gelfand–Levitan integral equation

$$F(r, t) + K(r, t) + \int_0^r dsK(r, s)F(s, t) = 0 ,$$

$$F(r, t) = \lim_{N \to \infty} \int_{-\infty}^N d\sigma(E) \cos(\sqrt{E}t) \cos(\sqrt{E}r) , \tag{2.285}$$

for the kernel K. The potential is then given by

$$V(r) = 2\frac{d}{dr}K(r, r) . \tag{2.286}$$

Uniqueness proof:

In order to solve the questions (i) and (iii) we propose to study the logarithmic derivative of $u(r, E)$ taken at the origin for a solution with suitable boundary conditions. First we make sure of the existence of such a solution.

Lemma:

Assume the existence of

$$\int_R^\infty dr \frac{V'^2}{V^{5/2}} , \quad \int_R^\infty dr \frac{V''}{V^{3/2}} \quad \text{for some } R ; \tag{2.287}$$

there exists a solution of $u'' = (V - E)u$ with asymptotic behaviour

$$u(r, E) \underset{r \to \infty}{\sim} \frac{\exp\left(-\int_R^r dr' \sqrt{V(r') - E}\right)}{(V - E)^{1/4}} = u_0(r, R) \tag{2.288}$$

for all complex values of E.

Proof:

$u_0(r, E)$ fulfils a Schrödinger equation with a modified potential. The Schrödinger equation for u can be transformed into an integral equation, the inhomogeneous term of which is precisely u_0. The finiteness of (2.287) allows the bounding of the resulting kernel, such that one can prove uniform convergence of the iteration procedure. Furthermore, $u(r, E)$ and $(d/dr)u(r, E)$ possess the analyticity properties of the individual terms of that series with respect to E. ∎

Theorem:

Assume $V(r)$ is locally integrable and fulfils (2.287). Then

$$R(E) = \frac{du(r, E)/dr}{u(r, E)}\bigg|_{r=0} \tag{2.289}$$

is a Herglotz function and admits the following representations:

$$-\frac{1}{R(E)} = \sum_{i=1}^{\infty} \frac{|u(0, E_i')|^2}{E_i' - E}, \quad \int_0^{\infty} dr u^2(r, E_i') = 1, \tag{2.290}$$

$$R(E) = C + E \sum_{i=l}^{\infty} \frac{|u'(0, E_i)|^2}{E_i(E_i - E)}, \quad \int_0^{\infty} dr u^2(r, E_i) = 1, \tag{2.291}$$

where E_i' is the unphysical and E_i the physical spectrum.

Proof:

Since we assumed V to be locally integrable (unfortunately that excludes the Coulomb case), integration down to $r = 0$ is allowed. For large values of r we use the above lemma, for small values we use Poincaré's theorem to conclude that $R(E)$ is a meromorphic function of E. By combining (2.282) with the complex conjugate and integrating, we get

$$\frac{\text{Im } R(E)}{\text{Im } E} = \frac{\int_0^{\infty} dr |u|^2}{|u(0)|^2} > 0. \tag{2.292}$$

This means $R(E)$ is a Herglotz function. *A priori*, such a function grows at most like $|E|$ and decreases at most like $1/|E|$ at ∞. It has poles and zeros only, interlaced on the real axis. The zeros give the unphysical spectrum E_i', the poles the physical one. To calculate the residues at the poles it is sufficient to construct the Wronskian of the solution at two nearby energies:

$$u'(0, E + \Delta E)u(0, E) - u'(0, E)u(0, E + \Delta E) = -\Delta E \int_0^{\infty} dr u(r, E)u(r, E + \Delta E). \tag{2.293}$$

For $E = E_i$ we get

$$\frac{d}{dE} R^{-1}(E)\bigg|_{E=E_i} = \frac{\int_0^{\infty} dr u^2(r, E_i)}{u'^2(0, E_i)} \tag{2.294}$$

and similarly for $E = E_i'$,

$$\frac{d}{dE}R(E)\bigg|_{E=E_i'} = -\frac{\int_0^\infty dr u^2(r, E_i')}{u^2(0, E_i')}.$$ (2.295)

■

To proceed we used the following lemma.

Lemma:
Let V be locally integrable; then the asymptotic behaviour of R is given as

$$|R(E) + \sqrt{-E}| \to 0 \quad E \to -\infty.$$

For a proof see Appendix A of Ref. [115]. The main idea is to show that the WKB approximation is valid for $E \to -\infty$. This was proved in two steps: first for V lower bounded, then for the general case.

Since $R(E)$ behaves like $-\sqrt{-E}$ for $E \to -\infty$, we can write an unsubtracted representation for $-1/R(E)$ (Eq. (2.290)) and a once-subtracted representation for $R(E)$ (Eq. (2.291)).

Remarks:
If we write (2.290) as a Stieltjes integral, the measure is exactly the spectral measure mentioned in (2.284). The fact that u/u' is entirely fixed from the E_i' and the $|u_i(0)|^2$ without subtraction explains why this case was treated first. There is a close analogy between the quantity we consider and the Wigner R matrix, which is $u(R)/u'(R)$ obtained by integrating from $r = 0$ up to the boundary of the nucleus, while we integrate the Schrödinger equation from ∞ down to zero.

This lemma shows that the subtraction constant is actually not free. Now we give an explicit procedure for obtaining it:

Lemma:
Let V be lower-bounded, monotonous for large r going to infinity faster than r^ϵ and slower than r^M. The constant in Eq. (2.291) is then given by

$$C = \lim_{N\to\infty} C_N \quad C_N = \sum_{i=1}^N \frac{|u'(0, E_i)|^2}{E_i} - \frac{2}{\pi}\sqrt{E_{N+1}'}.$$ (2.296)

Proof:
Let us mention that a semiclassical approximation for the wave functions and energy levels would immediately give (2.296). To prove (2.296), we computed $R(0)$ by computing a contour integral of $R(E)/E$ along

circles Γ_N of radius $|E'_{N+1}|$ centred at the origin containing the poles $E_1, ..., E_N$. This gives:

$$\frac{1}{2\pi i} \int_{\Gamma_N} \frac{dE\ R(E)}{E} = -\sum_{i=1}^{N} \frac{|u'(0, E_i)|^2}{E_i} + C . \qquad (2.297)$$

Next we prove that $R(E)$ is well approximated by $\sqrt{-E}$ for $|\arg(-E)| < \pi - \epsilon$, with ϵ small enough that the l.h.s. is close to $2\sqrt{E'_{N+1}}/\pi$ (see Ref. [115]). ∎

We have tested (2.296) by explicit calculations and found a very good agreement, $C \sim C_N$, even for small N.

Theorem:
Let V be locally integrable and fulfil the assumption of the last lemma; then the $\ell = 0$ bound-state energies E_i and the $|u'(0, E_i)|^2$ uniquely fix the potential.

Proof:
Taking the constants as stated fixes the subtraction constant (2.296); therefore $R(E)$ in (2.291) is determined. Then, taking the inverse (2.290) gives the spectral measure needed to solve the inverse problem according to Gasymov and Levitan. ∎

Remarks:
Since we know now that knowledge of E_i and $|u'_i(0)|^2$ also fixes the E'_i, it is enough to prove the convergence of the QRT procedure for the one-dimensional case for a symmetric potential in order to solve the radial problem.

Theorem:
Let V fulfil the assumptions of Theorem (2.289)–(2.291); let $\delta V(r)$ be a continuous function, the zeros of which have no accumulation points for finite r; then the E_i and $|u'_i(0)|$ uniquely determine the potential ($\delta V = 0$).

Proof:
Let $u(r, E)$ and $v(r, E)$ be the decreasing solutions of the Schrödinger equation for the potential, $V(r)$ and $V(r)+\delta V(r)$. We know from Theorem (2.282)–(2.286) that the same function $R(E)$ is associated with both u and v:

$$R(E) = u'(0, E)/u(0, E) = v'(0, E)/v(0, E) . \qquad (2.298)$$

Let r_1 be the first zero of $\delta V(r)$; then, combining the Wronskian, for u and v and using (2.298) gives

$$u(r_1, E)v'(r_1, E) - u'(r_1, E)v(r_1, E) = \int_0^{r_1} dr \delta V(r)u(r, E)v(r, E) . \quad (2.299)$$

For E large and negative we have, generalizing the last lemma to arbitrary r,

$$u(r, E)/u(0, E) = \exp(-\sqrt{-E}r)[1 + O(1)] \quad (2.300)$$

and a similar equation for v, as well as

$$u'(r, E)/u(r, E) \simeq -\sqrt{-E} + O(1) . \quad (2.301)$$

One gets from (2.299)

$$\int_0^{r_1} dr \delta V(r) \frac{u(r, E)v(r, E)}{u(0, E)v(0, E)} = O(\sqrt{-E}\exp(-2\sqrt{-E}r_1)) . \quad (2.302)$$

However, since δV, u and v have a constant sign, the l.h.s. of (2.302) is certainly larger than

$$\exp(-2\sqrt{-E}\alpha r_1) \int_0^{\alpha r_1} dr \delta V(r) \quad (2.303)$$

for any $0 < \alpha < 1$. Letting E go to $-\infty$ gives a contradiction, unless $\delta V = 0$ for $0 < r < r_1$, so we conclude that the l.h.s. of Eq. (2.299) is equal to zero and we can repeat the argument to prove that δV vanishes in $r_1 < r < r_2$ and so on.   ∎

Remark:
One can slightly weaken the assumption about δV. If $\int dr \delta V(r)$ has no accumulation points of zeros the argument holds because the estimates on u and v are also valid for higher derivatives.

Proposition:
If two potentials are finite for finite r, go to infinity like r^n, and have an infinite set of levels E_m in common such that $\sqrt{E_{m+1}} - \sqrt{E_m} \to 0$, then their difference δV cannot have compact support.

Proof:
Assume that δV has compact support $[0, R]$. This time integrate the Schrödinger equation from the origin with boundary conditions $u(0, E) = v(0, E) = 0$, and $u'(0, E) = v'(0, E) = 1$. This gives

$$u'(R, E)v(R, E) - v'(R, E)u(R, E) = -\int_0^R dr \delta V(r)u(r, E)v(r, E) = \Phi(E)$$
$$(2.304)$$

where $\Phi(E)$ is an entire function of order 1 of $k = \sqrt{(E)}$ and exponential type $2R$. It vanishes according to the assumption for $k = \pm i\sqrt{E_m}$, meaning on a set of infinite asymptotic density on the imaginary axis. According to a known theorem [120], it is identically zero. The product $u(r, E)v(r, E)$ can now be written according to the Paley–Wiener theorem as

$$\int_{-2r}^{2r} dx e^{ikx} w(r, x) . \tag{2.305}$$

Putting (2.305) into (2.304) we get

$$\int_0^R dr \delta V(r) w(r, x) = 0 , \tag{2.306}$$

which can be regarded as a Volterra equation for δV, because of the support properties of w, and hence $\delta V = 0$. ∎

From this proposition one deduces that changing a finite number of levels will produce a change in the potential which extends up to infinity. For that simple case the Gelfand–Levitan equation has a degenerate kernel and the explicit solution can be obtained by adapting the usual method to the case in hand.

Proposition:
Given a potential $V_1(r)$ with energies $E_j^{(1)}$ and normalization constants $\gamma_j^{(1)}$, a potential V_2 which has energy levels $E_i^{(2)}$ and constants $\gamma_i^{(2)}$ ($i = 1, ..., M$), instead of levels $E_j^{(1)}$ and constants $\gamma_j^{(1)}$ ($j = 1, ..., N$), is given by

$$V_2 = V_1 - 2 \left(\frac{d}{dr}\right)^2 \ln \mathrm{Det}\,(1 + A) , \tag{2.307}$$

where A is an $((n + m) \times (n + m))$-dimensional matrix with entries

$$A_{ij} = C_i \int_0^r dr' u_1(k_i, r') u_1(k_j, r') ,$$

$$C_i = \begin{cases} -\gamma_i^{(1)}, i = 1, ..., N \\ \gamma_i^{(2)}, i = N + 1, ..., N + M \end{cases}$$
$$k_i^2 = \begin{cases} E_i^{(1)}, i = 1, ..N \\ E_i^{(2)}, i = N + 1, ..., N + M \end{cases} \tag{2.308}$$

and u_1 denotes the regular wave function for potential V_1.

Proof:
Since one uses the standard 'artillery' of the inverse problem, we need only be very brief and sketch the steps. From the Wiener–Boas theorem

one deduces the 'Povzner–Levitan' representation for the regular wave functions u_1 and u_2 (corresponding to potentials V_1 and V_2), respectively,

$$u_i(k,r) = \frac{\sin kr}{k} + \int_0^r dr' K_i(r,r') \frac{\sin kr'}{k}. \tag{2.309}$$

Solving one equation for $\sin (kr)/k$ and introducing the solution into the other gives

$$u_2(E,r) = u_1(E,r) + \int_0^r dr' K(r,r') u_1(E,r'), \quad k^2 = E, \tag{2.310}$$

where the new kernel K is a function of K_1 and K_2 of (2.309); similarly one gets a reciprocal relation,

$$u_1(E,r) = u_2(E,r) + \int_0^r dr' \tilde{K}(r,r') u_2(E,r'). \tag{2.311}$$

Now multiplying (2.311) by $u_1(E,t)$, integrating with the spectral measure $d\rho_2(E)$ for potential V_2, and using the completeness property, we get

$$\int_{-\infty}^{+\infty} d\rho_2(E) u_2(E,t) u_1(E,r) = 0 \quad \text{for } t > r. \tag{2.312}$$

Note that the spectral measure used here corresponds to the 'right' boundary conditions at the origin $u_2(E_i,0) = 0$ and is therefore different from that used in (2.311). Then, multiplying (2.310) by $u_1(E,t)$, again integrating with $d\rho_2(E)$, and using (2.312) gives

$$\int_{-\infty}^{+\infty} d\rho_2(E) u_1(E,t) u_1(E,r) + \int_0^r dr' K(r,r')$$
$$\times \int_{-\infty}^{+\infty} d\rho_2(E) u_1(E,t) u_1(E,r') = 0, \quad t < r. \tag{2.313}$$

Adding the identity

$$K(r,t) = \int_{-\infty}^{+\infty} d\rho_1(E) \int_0^r K(r,s) u_1(E,s) u_1(E,t) ds, \quad r > t \tag{2.314}$$

to (2.313) leads us to the Gelfand–Levitan equation with kernel K:

$$K(r,t) + F(r,t) + \int_0^r dr' K(r,r') F(r',t) = 0, \quad r > t. \tag{2.315}$$

The input $F(r,t)$ is now given in terms of the spectral measure by

$$F(r,t) = \int_0^\infty d\sigma(E) u_1(E,r) u_1(E,t), \quad \sigma(E) = \rho_2(E) - \rho_1(E). \tag{2.316}$$

■

Since we intend to change a finite number of constants as specified

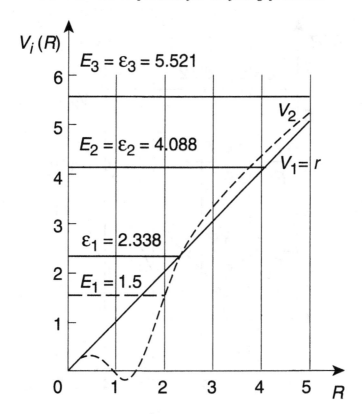

Fig. 2.4. The change in the potential when one starts from the linear potential and changes one level from 2.338 to 1.5.

earlier, σ is given by the finite sum:

$$\frac{d\sigma(E)}{dE} = \sum_{i=1}^{M} \gamma_i^{(2)} \delta(E - E_i^{(2)}) - \sum_{j=1}^{N} \gamma_j^{(1)} \delta(E - E_j^{(1)}) . \qquad (2.317)$$

For the case of a degenerate kernel it is well known that the Gelfand–Levitan equation admits the solution (2.307), (2.308) in closed form.

To illustrate this procedure, we have calculated δV, starting with the linear potential and changing one- and two-bound state levels (Figure 2.4 and 2.5 respectively).

Before turning to the inverse ℓ problem, let us remark that within the WKB approximation one already gets a unique monotonous potential given all energy levels.

Proposition:
Assume $V(r)$ is monotonically increasing so $r(V)$ exists and all energy levels are given. Within the WKB approximation we get

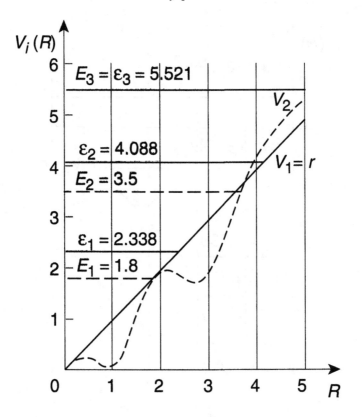

Fig. 2.5. The variation in the potential due to changing two levels.

(a) for the S-wave case [114]:

$$r(V) = 2 \int_{E_{min}}^{V} \frac{dE'}{\sqrt{V - E'}} \frac{dN(E)}{dE'} \; ; \qquad (2.318)$$

(b) for the three-dimensional case:

$$r^3(V) = 6 \int_{0}^{V} \frac{d^2 N(E')}{dE'^2} \frac{dE'}{\sqrt{V - E'}} . \qquad (2.319)$$

Proof:

(a) The number of bound states below some energy E is given in the semiclassical approximation by

$$N(E) = \frac{1}{\pi} \int_{0}^{E} dV \frac{dr}{dV} \sqrt{E - V} . \qquad (2.320)$$

Differentiating $N(E)$ once, with respect to E, we get an expression which can be regarded as an Abel integral equation for $r(V)$ and inverted; this gives (2.318).

(b) To proceed similarly in the three-dimensional case, we use the well-known formula for the total number of bound states below E:

$$N(E) \cong \frac{1}{6\pi^2} \int d^3x (E - V(r))^{3/2} . \tag{2.321}$$

It was first shown in Ref. [121] that (2.321) becomes exact in the strong coupling limit, where one takes $\lambda V(r)$ as a potential and takes the limit $\lambda \to \infty$. To invert (2.321), one differentiates twice and again gets an Abel integral equation:

$$\frac{d^2N}{dE^2} = \frac{1}{6\pi} \int \frac{dV \, dr^3(V)/dV}{(E - V)^{1/2}} . \tag{2.322}$$

Inverting (2.322) gives (2.319).

∎

Remark:

The obvious generalization of (2.321) to different 'moments' of sums over energy levels — or, rather, quantities like $\mathrm{Tr}\,\theta(-H)|H|^\alpha$ — has been discussed in the literature (see e.g. Ref. [127]: there also exist rigorous bounds of that type,

$$\sum_i |E_i|^\alpha \leq C_\alpha \int d^3x |V_-(x)|^{3/2+\alpha} , \tag{2.323}$$

where $|V_-|$ denotes the absolute value of the attractive part of V. Such bounds have led to the creation of a whole 'industry', since it was recognized that they are of great help in the problem of proving the 'stability of matter' (see Section 2.7)

The inverse problem for $E(\ell)$

We now want to discuss a non-standard inverse problem connected with confining potentials. Instead of giving ourselves the successive energy levels for a given angular momentum, we give ourselves the ground-state energies for all physical angular momenta, $E(\ell)$. The question is whether this uniquely fixes the potential or not. We shall not be able to give a complete answer to this question, but collect a certain number of indications towards the following conjecture:

Conjecture:

The confining potential is uniquely fixed by knowledge of the ground-state energies of the system for all angular momenta.

Example:

First, we look at a particularly simple case, that of equally spaced energy levels

$$E(\ell) = A + B\ell .$$ (2.324)

One solution of the problem is well known. It is the harmonic oscillator

$$V(r) = \frac{B^2}{4}r^2 + A - \frac{3B}{2} .$$ (2.325)

For this particular solution of the inverse problem, the reduced wave functions associated with $E(\ell)$ are given by (in the special case of $B = 2$)

$$u_\ell(r) = N_\ell r^{\ell+1}\exp(-r^2/2) .$$ (2.326)

We shall content ourselves with investigating the local uniqueness of V. Suppose we change the potential by δV, infinitesimal. Then by the Feynman–Hellmann theorem the change in the energy levels will be

$$\delta E_\ell = \int_0^\infty dr\delta V|u_\ell(r)|^2 .$$ (2.327)

An equivalent potential $V + \delta V$ will be such that all δE_ℓ vanish. The question is then one of completeness. From (2.326) and (2.327) we have, putting $r\delta V(r) = W(x), r^2 = x$,

$$\int_0^\infty dx W(x)x^\ell e^{-x} = 0, \quad \forall \ell \in \mathbf{N} .$$ (2.328)

According to Szegö [122] the system $\phi_\ell = x^\ell e^{-x}$ is 'closed', so that any function $f(x) = W(x)e^{-x}$, with a finite L^2 norm, can be approximated by $\sum c_\ell \phi_\ell$ in   such   a   way   that $\| f(x) - \sum_\ell c_\ell \phi_\ell \|_2 < \epsilon$, with ϵ arbitrarily small. That we need an assumption on $W(x)$ can be seen from observing that the system of powers $x^k, k = 0, 1, 2, ...$ is not complete with respect to a weight $\exp(-\ln^2|x|)$ [123], since

$$\int_0^\infty dx x^k e^{-\ln^2|x|} \sin(2\pi \ln|x|) = 0$$ (2.329)

for all k. This means that any function which has arbitrarily close mean-square approximations by means of polynomials for the above weight is orthogonal to $\sin(2\pi \ln|x|)$. This example is due to Stieltjes.

If δV grows more slowly than any exponential, $f(x)$ is guaranteed to have a finite L^2 norm and the theorem applies. Conditions (2.328) therefore guarantee that $\delta V \equiv 0$. Thus we get the following proposition:

Proposition:
If the levels $E(\ell)$ are equally spaced, the harmonic oscillator potential constitutes an isolated solution to the inverse problem.

If one tries to generalize this result to other potentials one first gets the set of equations

$$\int_0^\infty dr \delta V |u_\ell(r)|^2 = 0, \quad \forall \ell \in \mathbf{N} ;$$

however, one encounters a very difficult problem, which is to prove that the set of functions $|u_\ell(r)|^2$ is complete in some way; the mathematical literature is incredibly poor in this respect.

We shall nevertheless try to study these functions $|u_\ell(r)|^2$ and prove the following proposition:

Proposition:
Let $V(r)$ be a confining potential such that $V(r)/r^2 \to 0$ for $r \to \infty$. Then the reduced wave functions $u_\ell(r)$ associated with the levels $E(\ell)$, normalized by $\lim_{r\to 0} u_\ell(r)/r^{\ell+1} = 1$, are such that for any fixed R

$$\lim_{\ell\to\infty} \frac{u_\ell(R)}{R^{\ell+1}} = 1$$

and

$$\lim_{\ell\to\infty} \frac{u'_\ell(R)}{(\ell+1)R^\ell} = 1 .$$

Proof:
We incorporate the normalization condition by writing the integral equation

$$u_\ell(r) = r^{\ell+1} + \frac{1}{2\ell+1} \int_0^r dr' u_\ell(r') \left[\frac{r^{\ell+1}}{r'^\ell} - \frac{r'^{\ell+1}}{r^\ell} \right] [V(r') - E(\ell)] . \quad (2.330)$$

We define R_c to be the smallest r such that $E = V(R_c)$. If $r < R_c$ we have from Eq. (2.330) $u_\ell(r) < r^{\ell+1}$ and, by iteration,

$$u_\ell(r) > r^{\ell+1} \left[1 - \frac{E(\ell)}{2\ell+1} \int_0^r dr'r' \left(1 - \left(\frac{r'}{r}\right)^{2\ell+1} \right) \right], \quad (2.331)$$

and hence

$$r^{\ell+1} > u_\ell(r) > r^{\ell+1} \left[1 - \frac{E(\ell)}{2\ell+1} \frac{r^2}{2} \right] . \quad (2.332)$$

Since, as we have seen from the considerations in Section 2.3,

$$\frac{E(\ell)}{2\ell + 1} \to 0 \ \text{if} \ \frac{V(r)}{r^2} \to 0 \ \text{for} \ r \to \infty,$$

the first part of this proposition follows. The statement about the derivative of u_ℓ follows by differentiating (2.330). ∎

Remark:
We believe that the last proposition is indicative that the set of functions $u_\ell^2(r)$ is in some way complete, at least on a finite interval. According to a theorem of Müntz [124], a sequence of pure powers r^{α_n} is complete if the α_ns have a finite density for $n \to \infty$. However, even if we were to succeed in proving completeness over a finite interval, we would have the problem of letting the interval extend towards $+\infty$.

We shall, however, make another completely rigorous use of the last result, but restrict the generality of the possible variations of δV.

Theorem:
Assume that the set of levels $E(\ell)$ is reproduced by two potentials, U and V, such that U/r^2 and $V/r^2 \to 0$ for $r \to \infty$, and $U(r)$ and $V(r) \to \infty$ for $r \to \infty$. If, in addition, $U - V$ has a definite sign for $r > R$, then $U = V$ for $r > R$.

Proof :
Consider a given energy level $E(\ell)$. The wave functions, normalized as in the last proposition, associated with the potentials U and V are, respectively, u_ℓ and v_ℓ. By combination with the Schrödinger equation one gets

$$u_\ell' v_\ell - u_\ell v_\ell' \ |_R = \int_R^\infty (V - U) u_\ell v_\ell dr = \int_0^R dr u_\ell v_\ell (U - V).$$

Assume that for $r > R$, $V - U$ has a constant sign which can be chosen to be positive. We then have

$$u_\ell' v_\ell - u_\ell v_\ell' \ |_R > \int_{(1+\epsilon)R}^{(1+2\epsilon)R} dr u_\ell v_\ell (V - U), \quad \epsilon > 0. \tag{2.333}$$

For large ℓ the last proposition allows us to estimate both sides of (2.335). The l.h.s. is of the order of $(2\ell + 1)R^{2\ell+1}$, while the r.h.s. is larger than

$$(1 + \epsilon)^{2\ell+2} R^{2\ell+2} \int_{(1+\epsilon)R}^{(1+2\epsilon)R} dr (V - U),$$

which is clearly a contradiction, unless $V = U$ for $r > R$.

Now we are working with a compact interval and we have two possibilities, one being rigorous, the other imperfect. ∎

Corollary to the last theorem:
If, in addition, $V - U$ has a finite number of changes of sign, $V \equiv U$.

Proof:
First apply the theorem to R_n, corresponding to the last change of sign of $V - U$, then to R_{n-1}, etc. ∎

Pseudocorollary:
We keep the conditions of the last theorem. We admit that we have conditions sufficiently close to the Müntz theorem to believe that the set of the $u_\ell v_\ell$ is complete on a finite interval and conclude from $\int_0^R dr(U - V)u_\ell v_\ell = 0$ that $U \equiv V$.

This is as far as we have been able to go on a completely rigorous basis.

Now we shall describe a semiclassical treatment of the problem. Assume for simplicity $(d/dr)[r^3(dV/dr)] > 0$. This guarantees that $\ell(\ell+1)/r^2+V(r)$ has a unique minimum. It also implies that V is increasing.

We start by purely classical considerations, neglecting completely the radial kinetic energy. Then $E(\ell)$ will be the minimum of the centrifugal and the potential energies

$$E(\ell) = (\ell + 1/2)^2/r^2 + V(r), \tag{2.334}$$

with r given by

$$2(\ell + 1/2)^2/r^3 = dV/dr, \tag{2.335}$$

r being a function of ℓ. We can differentiate $E(\ell)$ with respect to ℓ, using an interpolation in ℓ, taking into account (2.333) and get

$$\frac{dE(\ell)}{d\ell} = \frac{(2\ell + 1)}{r^2} \rightarrow r^2 = \frac{2\ell + 1}{dE/d\ell}. \tag{2.336}$$

Remark:
r^2 can be rewritten as $(dE/d[\ell(\ell+1)])^{-1}$, which, according to the general concavity property, is an increasing function of ℓ.

With r being given as a function of ℓ, V can also be obtained as a function of ℓ, from (2.334)

$$V(\ell) = E(\ell) - \frac{1}{2}\left(\ell + \frac{1}{2}\right)dE/d\ell. \tag{2.337}$$

We see that at this level the inverse problem is incredibly simple. We shall try to see to what extent this kind of approach can be justified quantum mechanically. Interpolation in ℓ is not a problem. It was done long ago by Regge [125]. We shall return to this problem in the next section.

The quantum mechanical analogue of (2.334) is

$$E(\ell) = \int_0^\infty dr \left[u_\ell'^2 + \frac{\ell(\ell+1)}{r^2} u_\ell^2 + V(r) u_\ell^2 \right] , \qquad (2.338)$$

where from now on we take u_ℓ^2 to be normalized by $\int_0^\infty dr u_\ell^2 = 1$. Hence we get, using the well-known inequality $\int_0^\infty dr u^2/4r^2 \le \int_0^\infty dr u'^2$,

$$E(\ell) > \min \left(\left(\ell + \frac{1}{2} \right)^2 /r^2 + V(r) \right) \qquad (2.339)$$

which very much resembles (2.334), except for the fact that it is an inequality. The quantum mechanical analogue of (2.336) is

$$\frac{dE}{d\ell} = (2\ell + 1) \int_0^\infty dr \frac{u^2}{r^2} \qquad (2.340)$$

and, if the wave function is sufficiently concentrated around the minimum of the effective potential, reduces to (2.336). Our condition on the potential in fact allows us to make a very precise statement.

Proposition:

Define r_0 and r_1 by

$$\frac{1}{r_0^2} = \int_0^\infty dr \frac{u_\ell^2}{r^2}, \quad V(r_1) = \int_0^\infty dr u_\ell^2 V(r) \qquad (2.341)$$

and assume

$$\frac{d}{dr} \left(r^3 \frac{dV}{dr} \right) > 0,$$

then $r_1 > r_0$.

Proof:

We shall be satisfied by proving $r_1 \ne r_0$. If $r_1 = r_0$

$$\int_0^\infty dr u_\ell^2 \left\{ \frac{V(r) - V(r_0)}{V'(r_0)} + \frac{r_0^3}{2} \left(\frac{1}{r^2} - \frac{1}{r_0^2} \right) \right\} = 0 . \qquad (2.342)$$

Calling the curly bracket $y(r)$ we see that $y(r_0) = y'(r_0) = 0$, but $(d/dr)r^3$ $(dy/dr) > 0$, which guarantees $y(r) \ge 0$. The sign of the inequality between r_0 and r_1 is intuitively clear. ∎

Remark:

From the last proposition, we conclude that:

$$E(\ell) > V(r_0) + \frac{\ell + \frac{1}{2}}{2} \frac{dE}{d\ell}, \qquad r_0 = \sqrt{\frac{2\ell + 1}{dE/d\ell}}. \qquad (2.343)$$

This can be made more practical by using concavity in $\ell(\ell + 1)$ and avoiding the use of non-integer ℓ.

We have

$$\frac{2\ell + 1}{2\ell - 1}[E(\ell) - E(\ell - 1)] > \frac{dE}{d\ell} > \frac{2\ell + 1}{2\ell + 3}[E(\ell + 1) - E(\ell)] \qquad (2.344)$$

and hence we get the definition and the inequality

$$\bar{r}_0 = \sqrt{\frac{2\ell - 1}{E(\ell) - E(\ell - 1)}}, \qquad V(\bar{r}_0) < E(\ell) - \frac{1}{2}(\ell + \frac{1}{2})\frac{\ell + \frac{1}{2}}{\ell + \frac{3}{2}}[E(\ell + 1) - E(\ell)].$$
$$(2.345)$$

This is a strict and probably useful inequality, but it does not tell us how close we are to the potential.

Now we shall abandon rigour and try to approximate the effective potential $\ell(\ell + 1)/r^2 + V(r)$ by a harmonic potential and calculate the zero-point energy. To make this estimate, we use the potential determined without quantum effects and compute the curvature at the minimum. This gives

$$\begin{cases} V(r) = E(\ell) - \frac{\ell(\ell + 1)}{2\ell + 1}\frac{dE}{d\ell} - \sqrt{\frac{V''}{2} + 3\frac{\ell(\ell+1)}{r^4}} \\[2mm] r^2 = \frac{2\ell + 1}{dE/d\ell} \\[2mm] V'' = -\frac{1}{2}\left(\frac{dE}{d\ell}\right)^2 + \left\{\frac{d}{d\ell}\left[(2\ell + 1)\left(\frac{dE}{d\ell}\right)\right]\right\} / \left\{\frac{d}{d\ell}\left[(2\ell + 1)/\left(\frac{dE}{d\ell}\right)\right]\right\}. \end{cases} \qquad (2.346)$$

Unfortunately, it seems difficult to find a systematic iterative procedure to improve this formula. As it stands, it gives rather good results for large r.

Example:

Take $E_0 = 2\ell + 3$ ($V = r^2$). Equations (2.343) give $V = r^2 + 1$, while Eqs. (2.346) give

$$V(r) = 2r^2 + 2 - \frac{r^{4 - \frac{1}{4}}}{r^2} - \sqrt{1 + \frac{3(r^4 - \frac{1}{4})}{r^4}} \simeq r^2 + \frac{1}{4r^2} + \dots$$

and therefore there is no constant term.

Example:
Let

$$E = -\frac{1}{4(\ell+1)^2} \quad \left(V = -\frac{1}{r}\right).$$

Equations (2.346) give

$$V = -1/r + O(1/r^2),$$

while (2.343) contains, after the leading term, a contribution in $r^{-3/2}$.

Our conclusion is that empirically our improved semiclassical procedure is excellent for obtaining the large-distance behaviour of the potential, and can certainly be used as a starting point in the search for the exact potential. On the other hand, a very well-posed, but difficult, mathematical problem remains.

Regge trajectories for confining potentials

We have previously been led to introduce an interpretation of the ground-state energies for non-integer ℓ. In Section 2.1 we also mentioned convexity properties in ℓ. Confining potentials are in a way simpler than ordinary potentials in regard to analyticity properties in the angular momentum complex plane, because of the absence of the continuous spectrum.

Following Regge [125] we can try to solve the Schrödinger equation for complex ℓ, with Re $\ell > -1/2$:

$$\left(-\left(\frac{d}{dr}\right)^2 + \frac{\lambda}{r^2} + V(r)\right) u(r) = E(\ell)u(r), \quad \lambda = \ell(\ell+1), \quad u(0) = 0.$$
$$(2.347)$$

By multiplication by u^* and integration we immediately get

$$\int_0^\infty dr \left\{|u'|^2 + \frac{\ell(\ell+1)}{r^2}|u|^2 + V(r)|u|^2\right\} = E(\ell)\int_0^\infty dr|u|^2, \quad (2.348)$$

which implies that normalizable solutions of the Schrödinger equation are such that

$$\text{Im } \lambda \int_0^\infty dr\frac{|u|^2}{r^2} = \text{Im } E \int_0^\infty dr|u|^2, \quad (2.349)$$

and hence for real angular momentum E is real. In non-confining, rapidly decreasing potentials, solutions of the Schrödinger equation for ℓ real and large enough, are no longer normalizable, the reality argument breaks down and, as everybody knows, the energies become complex or, conversely, real positive energies correspond to complex angular momenta.

We think it is worth studying Regge trajectories in confining potentials, because there are practically no examples of Regge trajectories known, except in the oscillator and Coulomb cases, and for numerical calculations in the Yukawa potential.

The problem is that $E(\ell)$, defined implicitly by (2.347), is analytic whenever the derivative

$$\frac{dE}{d\ell} = (2\ell + 1) \int_0^\infty dr \frac{u_\ell^2}{r^2} / \int_0^\infty dr u_\ell^2 \qquad (2.350)$$

exists, but except for ℓ real nobody knows when this derivative exists because $\int_0^\infty dr u_\ell^2$ may vanish. What is known is that if $E(\ell)$ — or $\ell(E)$ — has isolated singularities in the complex ℓ plane, these singularities cannot be poles or essential singularities because from (2.349) Im $E/$Im $\lambda > 0$, or since Re $\ell > -1/2$, Im $E(\ell)/$Im $\ell > 0$.

From now on we shall restrict ourselves to the pure power potentials

$$V(r) = r^\alpha . \qquad (2.351)$$

This will allow us to use the complex r plane. In fact, most of what follows is an adaptation of the work of Loeffel and one of the present authors (A.M.) [81, 126] on the anharmonic oscillator.

For a potential of the type (2.351) the solution of the Schrödinger equation behaves at large distances like the WKB solution. One can use the WKB solution as a starting point to establish an asymptotic expansion of the solution:

$$u_0 = r^{-\alpha/4} \exp\left\{-\frac{r^{\alpha/2+1}}{\alpha/2+1}\right\} (1 + O(r^{1-\alpha/2})) \quad \text{for } \alpha > 2, \qquad (2.352)$$

$$u_0 = r^{\ell+1} \exp\left(-\frac{r^2}{2}\right) \quad \text{for } \alpha = 2, \qquad (2.353)$$

$$u_0 = r^{-\alpha/4} \exp\left\{-\frac{r^{\alpha/2+1}}{\alpha/2+1} + \frac{E}{2} \frac{r^{1-\alpha/2}}{1-\alpha/2} + O(r^{1-3\alpha/2})\right\} \quad \text{for } \alpha < 2. \qquad (2.354)$$

In the case $\alpha < 2$, more and more terms are needed in the expansion as α approaches zero, if one wants u_0 to be such that $\lim_{r\to\infty}(u/u_0) = 1$.

Because the potential is an analytic function of $z = r$ except at the origin, the solution can be analytically continued in the complex r plane for fixed ℓ (real or complex). It can be shown that the solution is decreasing in the sector

$$-\frac{\pi}{2+\alpha} < \arg z < \frac{\pi}{2+\alpha} ,$$

where it behaves like the analytic continuation of (2.352)–(2.354) — i.e., it is dominated by $\exp\{-z^{(\alpha/2+1)}/(\alpha/2+1)\}$.

Remark (a):

If $\alpha > 2$, the two independent solutions behave like

$$z^{-\alpha/4} \exp\left\{\pm\frac{z^{\alpha/2+1}}{(\alpha/2)+1}\right\}$$

and hence $|u|^2$ is still square integrable along the rays $\arg z = \pm\pi/(2+\alpha)$.

We shall need to continue the solution in $-3\pi/(\alpha+2) < \arg z < 3\pi/(\alpha+2)$ further. In the sector $\pi/(\alpha+2) < \arg z < 3\pi/(\alpha+2)$, u is, *a priori*, a superposition of solutions asymptotically behaving like $u_0 = z^{-\alpha/4} \exp(P(z))$, $v_0 = z^{-\alpha/4} \exp(-P(z))$, where $P(z) = -(z^{\alpha/2+1}/(\alpha/2 + 1)) + (E/2)(z^{1-\alpha/2}/(1-\alpha/2)) + \dots$ contains only positive powers of z. But $u_0/v_0 \to \infty$ for $\pi/(\alpha+2) < \arg z < 3\pi/(\alpha+2)$ and hence $|u/u_0| < \text{const}$. We get the following situation: $u/u_0 \to 1$ for $|\arg z| < \pi/(2+\alpha)$, $|u/u_0| < \text{const}$ for $|\arg z| < \min\{3\pi/(2+\alpha), \pi\}$. According to Montel's theorem we conclude that

$$u/u_0 \to 1 \quad \text{for } |\arg z| < \min\{3\pi/(2+\alpha), \pi\}. \qquad (2.355)$$

Let us now describe our strategy for establishing the analyticity properties of the energy levels $E_n(\ell)$ with respect to ℓ or λ. We have seen already that the only singularities to worry about are branch points (natural boundaries can be discarded after a difficult argument which we shall not reproduce here). To exclude branch points, first find a characterization of the successive energy levels for complex ℓ. For ℓ real $> -1/2$ the levels are labelled by the number of zeros of $u(r)$ on the real positive axis. The first step will be to show that in the angle

$$|\arg z| < \min\{\pi, 3\pi/(2+\alpha)\}$$

there are no other zeros for real ℓ. We can show that it is still possible to characterize these energy levels for complex ℓ by the number of zeros either in $|\arg z| < \pi/(2+\alpha)$ in the case in which $\alpha > 2$, or in

$$-\frac{\pi}{2+\alpha} < \arg z < \frac{2\pi}{2+\alpha} \quad \text{for } \alpha < 2, \quad \text{Im } \lambda > 0$$

and

$$-\frac{2\pi}{2+\alpha} < \arg z < \frac{\pi}{2+\alpha} \quad \text{for } \alpha < 2, \quad \text{Im } \lambda < 0.$$

We shall prove that the number of zeros in these regions does not vary along any path in the complex ℓ plane starting from and returning to the real axis, with Re $\ell > -1/2$. This will demonstrate the absence of branch points.

Step I:

ℓ real $> -1/2$. The Schrödinger equation written along a complex ray $z = t\, e^{i\phi}$ becomes

$$\left(-\left(\frac{d}{dt}\right)^2 + \frac{\lambda}{t^2} + t^\alpha e^{i(2+\alpha)\phi}\right) u = e^{2i\phi} E u . \qquad (2.356)$$

$$\operatorname{Re} u'u^* = \int_0^t dt' \left\{|u'|^2 + |u|^2 \left(\frac{\lambda}{t'^2} + t'^\alpha \cos(2+\alpha)\phi - E\cos 2\phi\right)\right\}, \qquad (2.357)$$

$$\operatorname{Im} u'u^* = \int_0^t dt'\{t'^\alpha \sin(2+\alpha)\phi - E\sin 2\phi\}|u|^2 . \qquad (2.358)$$

If $|\phi| < \pi/(2+\alpha), u \to 0$ at infinity and the integration sign $\int_0^t$ in (2.357) and (2.358) can be replaced by $-\int_t^\infty$. In this case we see that since the integrand in (2.358) is monotonous, we can always choose the limits in such a way that the integrand has a constant sign. Therefore, we have the following:

Property 1:

For ℓ real $> -1/2$, u and u' have no zero in

$$0 < |\phi| < \pi/(2+\alpha) .$$

We try now to investigate $\pi/(2+\alpha) \le |\arg \phi|$. One considers the combination

$$\operatorname{Re}(u'u^*)\cos \beta - \sin \beta \operatorname{Im}(u'u^*) .$$

This will be different from zero if one can find β such that

$$\cos \beta > 0, \quad \cos((2+\alpha)\phi + \beta) > 0, \quad \cos(2\phi + \beta) < 0 .$$

This is possible, if simultaneously

$$|\phi| < 3\pi/(\alpha+2), \quad |\phi| < 2\pi/\alpha .$$

Hence we get:

Property 2:

For ℓ real $> -1/2$ and $\alpha \le 4$, u and u' have no zeros in

$$0 < |\phi| < \min(3\pi/(\alpha+2), \pi) .$$

Step II:
We make λ complex, for instance, Im $\lambda > 0$. Equation (2.356) then becomes

$$\text{Im } u'u^* = \int_0^t dt'|u|^2 \left\{ \frac{\text{Im } \lambda}{t'^2} + t'^\alpha \sin(2+\alpha)\phi - |E|\sin(2\phi + \arg E) \right\} .$$
$$(2.359)$$

If $0 < \phi < \pi/(2+\alpha)$ the integral tends to zero if $t \to \infty$. However, $\sin(2+\alpha)\phi > 0$; so we conclude $\sin(2\phi + \arg E) > 0$. Hence, using continuity we get:

$$0 < \arg E < \alpha\pi/(2+\alpha), \quad \text{Im } \lambda > 0 . \tag{2.360}$$

Next, we prove a number of properties of zeros.

Property 3:
For Im $\lambda > 0$, u has no zeros for $-\pi/(2+\alpha) < \phi \le 0$.

Proof:
For ϕ in that interval the integrand in (2.359) is *monotonous* and one can choose the integration limits as $(0, t)$ or (∞, t). ∎

Property 4:
For Im $\lambda > 0$, $\alpha > 2$, u has no zero for

$$\phi = \pi/(2+\alpha) .$$

Proof:
According to Remark (a), $|u|$ is square integrable and we get

$$\text{Im } u'u^* = \int_0^t dt'|u|^2 \left\{ \frac{\text{Im } \lambda}{t'^2} - |E|\sin\left(\frac{2\pi}{\alpha+2} + \arg E\right) \right\} = -\int_t^\infty dt'|u|^2\{...\} .$$
$$(2.361)$$

Again, since the integrand is monotonous Im $u'u^*$ cannot vanish. ∎

Property 5:
For Im $\lambda > 0$, $\alpha < 2$, u has no zeros for

$$\phi = 2\pi/(2+\alpha) .$$

Proof:
Here we can only use the limits $(0, t)$:

$$\text{Im } u'u^* = \int_0^t dt'|u|^2 \left\{ \frac{\text{Im } \lambda}{|t'|^2} - |E|\sin\left(\frac{4\pi}{2+\alpha} + \arg E\right) \right\} . \tag{2.362}$$

According to (2.360) and condition $\alpha < 2$, we have

$$\pi < 4\pi/(2+\alpha) < \frac{4\pi}{(2+\alpha)} + \arg E < (4+\alpha)\pi/(2+\alpha) < 2\pi ,$$

and the integrand in the above expression is positive. ∎

Finally we notice:

Property 6:
u has no zeros for $|t|$ large enough, and

$$|\phi| < \min\{\pi, 3\pi/(2+\alpha)\} .$$

Proof:
This follows from the remark that $u/u_0 \to 1$ in this sector and that u_0 has no zeros. ∎

Step III:
Continuation in λ: We vary λ continuously, starting from λ real $> -1/4$ and taking for instance the case of Im $\lambda > 0$. Initially, from Property 6 we know that in the angle $|\phi| < \min[3\pi/(2+\alpha), \pi]$ if $\alpha < 4, |\phi| < \pi/(2+\alpha)$ if $\alpha > 2$, $u(r)$ has only zeros on the positive real axis, $\phi = 0$, the number of zeros (not including $z = 0$) being equal to $n-1$ for the n-th level.

Now we distinguish two cases:

(i) $\alpha > 2$. If we make λ complex there cannot be any zero on the lines $\phi = \pm\pi/(\alpha+2)$ from Properties 3 and 4 and no zero can come from infinity in the sector $|\phi| < \pi/(\alpha+2)$, from Property 6. Since the zeros are a continuous function of λ, their number inside the sector $|\phi| < \pi/(\alpha+2)$ is equal to what it was initially, for Re λ. If we return to Re λ, we get back to the *same* energy level, since the number of zeros has not changed. Therefore, there are no branch points in the λ complex plane cut from $-\infty$ to $-1/4$.

(ii) $\alpha < 2$. Take, for instance, Im $\lambda > 0$. There are no zeros in $-\pi/(\alpha+2) < \phi \leq 0$, no zeros on the line $\phi = 2\pi/(2+\alpha)$ and no zeros at infinity in the sector $0 < \phi < \min\{3\pi/(2+\alpha), \pi\}$. Therefore, again, the number of zeros in the sector $-\pi/(\alpha+2) < \phi < 2\pi/(\alpha+2)$ remains the same as for Re λ and continues to characterize the energy levels. Again we conclude that $E(\lambda)$ has no branch points in the λ complex plane, cut from $-\infty$ to $-1/4$.

We conclude with:

Theorem:

The energy levels $E_n(\ell)$ in a potential $V = r^\alpha$ can be continued for complex ℓ. The functions $E_n(\ell)$ are analytic in Re $\ell > -1/2$.

Remarks:

We also have the property

$$\text{Im } E_n(\lambda = \ell(\ell+1))/\text{Im } \lambda > 0 \,,$$

which indicates that $E_n(\lambda)$ is a Herglotz function, which can be written as

$$E_n(\lambda) = A + B\lambda + \frac{\lambda}{\pi} \int_{-\infty}^{-1/4} \frac{d\lambda' \text{Im } E_n(\lambda')}{\lambda'(\lambda' - \lambda)} \,, \qquad (2.363)$$

with $\text{Im } E_n > 0$. It is easy to check that this fits with the concavity of E_n with respect to $\ell(\ell+1)$ mentioned in Section 2.1.

However, we have

$$|\arg E| < \alpha\pi/(2+\alpha) \,.$$

If $\alpha < 2$, this means $|\arg E| < \pi/2$. We conclude that $[E_n(\lambda)]^2$ is also a Herglotz function, and therefore $E_n(\lambda)$ increases, at most, like $\lambda^{1/2}$ if $\alpha \le 2$. Since $(\lambda + 1/4)^{1/2} = \ell + 1/2$, it is tempting to speculate that for $\alpha \le 2$, $E_n(\ell)$ is itself a Herglotz function. However, we would need to establish the analyticity of $E_n(\ell)$ for Re $\ell < -1/2$, and this is difficult. The Herglotz property of $E_n(\ell)$ for $\alpha < 2$ fits with our proposition on the concavity of $E_n(\ell)$ with respect to ℓ.

More generally, $[E_n(\lambda)]^{(2+\alpha)/\alpha}$ is a Herglotz function, and therefore $E_n(\lambda)$ cannot grow faster than $\ell^{2\alpha/(\alpha+2)}$, for large ℓ. This fits with Equation (2.82). One can also exploit the fact that a Herglotz function is concave, and get useful bounds for arbitrary n.

Conclusion

It would be very desirable to extend these results to superpositions of powers, at least with positive weights. However, it is a weakness of our approach that it cannot handle these superpositions, because the various critical lines in the z complex plane change with α.

2.7 Counting the number of bound states

Motivation — History

Many-body problems are not easy to treat in general. But at least some questions can be handled by studying related non-linear one-body problems. This applies especially to the so-called *'stability of matter'* problem. The question is: Why is a system of N particles interacting via Coulomb forces stable [5, 6]? Or, expressed differently: Does the ground-state energy per particle converge to a finite limit for $N \to \infty$? Let us first quote the many-body Hamiltonian

$$H_N = \sum_{j=1}^{N} \frac{\vec{p}_j^2}{2m} + \sum_{i<j}^{N} \frac{e_i e_j}{|x_i - x_j|} . \qquad (2.364)$$

We are interested in the fermionic ground-state energy E_N, corresponding to a totally antisymmetric wave function with spatial part $\psi(\vec{x}_1, \ldots, \vec{x}_N)$. A very rough simplified argument would be the following: If N fermions occupy a volume V, the volume available for a particle is of the order of V/N; let V be a cube of side length R. A characteristic length is therefore of the order of $R/N^{1/3}$. From the uncertainty principle we would estimate the kinetic energy as

$$N \cdot \langle p_i^2 \rangle \simeq N \frac{N^{2/3}}{R^2} = \rho^{5/3} R^3 , \qquad (2.365)$$

where we have introduced the density $\rho = N/R^3$ which becomes in the Thomas–Fermi model the one-particle density

$$\rho(x_1) = \int d^3 x_2 \ldots d^3 x_N |\psi(\vec{x}_1, \vec{x}_2, \ldots, \vec{x}_N)|^2 . \qquad (2.366)$$

In fact, in the Thomas–Fermi model the kinetic energy is replaced by the $5/3$ moment of ρ times a suitable constant:

$$T(\psi) = \left\langle \psi \left| \sum_{j=1}^{N} \frac{\vec{p}_j^2}{2m} \right| \psi \right\rangle \geq c \int d^3 x \rho^{5/3}(x) . \qquad (2.367)$$

In order to derive (2.367) for an arbitrary antisymmetric ψ, we start from the Hamiltonian $h_N = \sum_{j=1}^{N} h_j$, where $h_j = \vec{p}_j^2/2m + V(x_j)$ and the one-particle potential is chosen to be $V(x) = -\lambda \rho^{2/3}(x)$. From the variational principle we deduce an upper bound on the fermionic ground-state energy E_0^F; a lower bound is obtained from filling the one-particle energy levels

$$- \kappa \int |V(x)|^{5/2} d^3 x \leq \sum_i e_i \leq E_0^F \leq \langle \psi | H_0 | \psi \rangle = T(\psi) - \lambda \int \rho^{5/3}(x) d^3 x. \qquad (2.368)$$

The first inequality of (2.368) allows us to deduce (2.367). This is a crucial step that has led to a whole industry and many attempts have been made to obtain bounds of this type. More generally, we may ask for bounds on the moments of energy levels which are of the quasiclassical phase-space integrals type [127]

$$\sum_i |e_i|^\alpha \cong \int \frac{d^d x \, d^d p}{(2\pi)^d} |(p^2 + \lambda V)_-|^\alpha = C\ell_{\alpha,d} \int d^d x |\lambda V_-(x)|^{d/2+\alpha}, \quad (2.369)$$

$$C\ell_{\alpha,d} = \frac{\Gamma(\alpha+1)}{(4\pi)^{d/2}\Gamma\left(\alpha+1+(d/2)\right)}.$$

In (2.369) we sum over all negative-energy eigenvalues. $|V_-(x)|$ denotes the absolute value of the attractive part of $V(x)$. A simple shift allows us to evaluate the energy levels below a fixed energy.

We note that the quasiclassical limit $\hbar \to 0$ is related by scaling to the strong coupling limit. The first proof that the number of bound states behaves as given by (2.367) in one-dimensional Schrödinger problems is due to Chadan [128]. The general d-dimensional case has been studied by one of the authors (A.M.) [129]. A further remark concerns relations between different moments. Let us denote by $N_\mathscr{E}$ the total number of bound states below an energy value $\mathscr{E}$. We obtain

$$\sum_j |e_j|^\alpha = \int_{-\infty}^0 dN_\mathscr{E} |\mathscr{E}|^\alpha = -\int_0^\infty dN_{-\mathscr{E}} \mathscr{E}^\alpha = \alpha \int_0^\infty d\mathscr{E} \cdot \mathscr{E}^{\alpha-1} N_{-\mathscr{E}} \quad (2.370)$$

and relate an integral over the zero moment to the α moment.

We should remark that for a complete solution of the stability problem an estimate of the potential energy is also needed. The simple-minded argument would run as follows: Positive and negative charges may be arranged in an alternating way and the averaged distance between nearest neighbours will be of the order of $R/N^{1/3}$. The total energy will then behave as

$$E_N \simeq \frac{N^{5/3}}{R^2} - \frac{N^{4/3}}{R}, \quad (2.371)$$

which, optimized with respect to R, yields E_N, proportional to N. But the estimate of the potential energy is delicate, since $N(N-1)/2$ terms contribute to (2.364). There is an electrostatic inequality,

$$\sum_{i<j}^N \frac{1}{|\vec{x}_i - \vec{x}_j|} \geq \sum_i^N \int \frac{d^3 x \rho(x)}{|\vec{x} - \vec{x}_i|} - \frac{1}{2} \int d^3 x \, d^3 y \frac{\rho(x)\rho(y)}{|\vec{x} - \vec{y}|}$$

$$-c_1 \int d^3 x \rho^{5/3}(x) - \frac{N}{2} c_2, \quad \text{with} \int d^3 x \rho(x) = N, \quad (2.372)$$

(with c_1 and c_2 suitable constants), which allows binding of the potential energy from below. We shall not go into the technical details but just say that together (2.368) and (2.372) allow us to prove the stability result. An improved bound of the type (2.368) will also improve the stability bound.

The history of estimates of the type (2.369) can be traced back to a problem treated by Weyl in 1911: We take as an operator the negative Laplacian $-\Delta_\Omega$ with Dirichlet boundary conditions on a domain $\Omega \subset \mathbf{R}^d$. The number of frequencies $N_d(\Omega, \lambda)$ below the value λ grows with λ as

$$N_d(\Omega, \lambda) \overset{\lambda \to \infty}{\cong} \lambda^{d/2} |\Omega| , \qquad (2.373)$$

where $|\Omega|$ denotes the volume of the domain Ω.

A number of semiclassical expansions have been treated in the literature and have become famous. Kac asked whether one can hear the shape of a drum. A simple argument shows that frequencies λ_m proportional to m^2 must come from a circular drum. We expand the trace of the heat kernel of $-\Delta_\Omega$ as

$$\mathrm{Tr}\ e^{t\Delta_\Omega} = \sum_m e^{-\lambda_m t} \overset{t \searrow 0}{\cong} \frac{c_0 |\Omega|}{t} + \frac{c_1 L}{t^{1/2}} + c_2 + \dots , \qquad (2.374)$$

with known constants $c_0, c_1, c_2, \dots$. Here $|\Omega|$ denotes the area of the drum, L denotes the length of the boundary of the two-dimensional domain, while c_2 is determined by the Euler characteristic: If all frequencies are known, the number of holes in the drum is determined. $|\Omega|$ and L are in general related by an isoperimetric inequality $|\Omega| \leq L^2/4\pi$. This inequality becomes an equality if the drum is circular. This essentially proves the uniqueness of the inverse problem for the case of a circular domain. The general inverse problem is much more complicated and has not been solved in general.

The analogue of expansion (2.374) for potential problems in one dimension deserves mention, too: For reasons of simplicity we treat $H = -d^2/dx^2 + V(x)$ defined on a finite interval $[0, L]$, so that H has a discrete spectrum. The expansion becomes

$$\mathrm{Tr}\ e^{-tH} \overset{t \searrow 0}{\cong} \frac{1}{\sqrt{t}} \sum_{n=0}^{\infty} c_n t^n I_n , \qquad (2.375)$$

where I_n are functionals of $V(x)$ and derivatives of $V(x)$: $I_1 = \int_0^L dx V(x)$, $I_2 = \int_0^L dx V^2(x)$, $I_3 = \int_0^L dx (V^3 + V_x^2), \dots$ and known constants c_n. Since the spectrum of H is left invariant under the KdV-flow, the functionals I_n are invariant, too, and are related to the higher conserved quantities of the KdV equation.

We may add an amusing remark about the stability of the hydrogen atom. We learn in lectures that the uncertainty principle 'proves' that the hydrogen atom is stable. For $h = -\Delta - \alpha/r$ the infimum of the spectrum is finite. The uncertainty principle shows only that the product of the mean square deviations of the momentum operator Δp times the mean square deviation of the position operator Δx is lower-bounded

$$\Delta p \Delta x \geq \frac{3\hbar}{2}, \qquad (2.376)$$

and it is amusing to note that there is no way to show, using only (2.376), that $\langle \psi | H \psi \rangle$ is bounded from below. A sequence of trial functions which respect (2.376), but which lead to expectation values of H that tend to $-\infty$ is easily constructed. We take a wave packet which is supported 50% on a big sphere of radius R and 50% on a small sphere of radius ε; then $\langle x^2 \rangle \simeq R^2/2$ but $\langle 1/r \rangle \simeq 1/2\varepsilon$. The expectation value of $\langle H \rangle \simeq 2/R^2 - \alpha/2\varepsilon$ will tend to $-\infty$ for ε going to zero, and ε and R are unrelated. This shows that we need a 'better uncertainty principle' which allows stability to be shown. The inequality due to Sobolev from 1938 allows one to do so. It allows binding of the kinetic energy by a certain moment of the density

$$s^{2/3} \int d^3x |\vec{\nabla}\psi(x)|^2 \geq \left(\int d^3x |\psi(x)|^6 \right)^{1/3}, \quad s = \frac{4}{3\sqrt{3\pi^2}} \simeq 0.0780. \quad (2.377)$$

More details concerning (2.377) and the evaluation of the Sobolev constant s will be given in Appendix C. Here, we cite that the use of (2.377) allows us to obtain a finite lower bound to the ground-state energy of the hydrogen atom

$$\inf \left\langle \psi \left| \left(-\Delta - \frac{\alpha}{r} \right) \psi \right\rangle \geq \inf \left\{ s^{-2/3} \left(\int |\psi|^6 \right)^{1/3} - \alpha \int \frac{|\psi|^2}{r} \right\} \quad (2.378)$$

which is finite. The actual value is -1.3 Ry, which differs from the exact answer by 30 per cent.

A non-trivial lower bound on $H = -\Delta + V(\vec{x})$ for potentials which are in a L^p-class for $p > 3/2$ has been obtained by one of the authors (H.G.):

$$H = -\Delta + V \geq -c_p \|V\|_p^{2p/(2p-3)}, \qquad p > 3/2, \quad (2.379)$$

where the constant c_p is given by

$$c_p = \Gamma \left(\frac{2p-3}{p-1} \right)^{\frac{2p-2}{2p-3}} \left(\frac{p-1}{p} \right)^2 (4\pi)^{-\frac{2}{2p-3}}.$$

The bound (2.379) indicates that the behaviour of a potential at the origin like $V(r) \overset{r \searrow 0}{\simeq} -1/r^{(2-\varepsilon)}$ with $\varepsilon > 0$ gives a Hamiltonian which is lower-bounded.

A combination of the argument leading to (2.377) and Hölder's inequality leads to a criterion which allows the exclusion of bound states [130].

The energy functional may be bounded as

$$E(\psi) = \frac{\int d^3x(|\vec{\nabla}\psi|^2 + V(x)|\psi(x)|^2)}{\int d^3x|\psi(x)|^2} \geq$$

$$\frac{(\int d^3x|\psi|^6)^{1/3}}{\int d^3x|\psi|^2}\left(s^{-2/3} - (\int d^3x|V_-(x)|^{3/2})^{2/3}\right). \qquad (2.380)$$

Therefore, if $s \int d^3x|V_-|^{3/2}(x)$ is less than one, bound states are excluded.

Locating bound states

There exist a few general results which help us to obtain insights into a number of situations.

For example, let us assume that we have two potentials, V_1 and V_2, which are related such that $V_1 \geq V_2$ for all $\tilde{x}$. Then the number of bound states below some energy E for the first problem cannot exceed the number of bound states for the second case: $N_E(V_1) < N_E(V_2)$. This follows from the min-max principle (Eq. (2.7)) by taking the true eigenfunctions for the first case as trial functions for the second. As an application we may compare a potential problem with potential $V_1(\tilde{x}) = V(\tilde{x})$ to another one where only the attractive part of V is taken into account $V_2(\tilde{x}) = -|V_-(\tilde{x})| = +V(\tilde{x})\theta(-V)$. Then clearly, $V_2 \leq V_1$ and $N_E(V) \leq N_E(-|V_-|)$.

Another example is given by taking V_1 to be some spherical, symmetric, attractive potential up to a radius R and setting V_1 outside infinite, which means Dirichlet boundary conditions. V_2 is taken to be identical to V_1 up to R, but constant, which we may put to zero, outside. Then, if n_0 is the number of bound states with angular momentum zero, $n_0(V_1) < n_0(V_2)$, but $n_0(V_2) - n_0(V_1) \leq 1$, too — since the number of bound states equals the number of nodes of the wave function for the appropriate energy.

There are a number of ways to estimate how many bound states may occur. It depends very much on the dimensionality. A few examples follow.

We have already mentioned that any one-dimensional Schrödinger operator $-\Delta + V(x)$ such that $\int dx V(x)$ is negative has at least one bound state. This can be shown by taking a Gaussian trial function $\varphi_\lambda = \lambda^{1/2}e^{-\lambda x^2}$ and observing the scaling behaviour of $\langle\varphi_\lambda|\Delta\varphi_\lambda\rangle$ and $\langle\varphi_\lambda|V\varphi_\lambda\rangle$. The condition on the integral over the potential is necessary. A counterexample with $\int_{-\infty}^{\infty} dx V(x) > 0$, which does not have a bound state, is given by a sum of two δ-potentials $V(x) = -\lambda\delta(x+R) + \mu\delta(x-R)$, $\lambda > 0$, $\mu > 0$, but $\mu > \lambda$. If λ is small enough (depending on R) no bound state occurs.

A similar result holds in two dimensions. Any two-dimensional Schrödinger operator $-\Delta + V$ with $\int dx \int dy V(x,y) < \infty$ has at least one bound state. We take again a trial function, this time in the form $e^{-\lambda r^\alpha}$. We obtain an upper bound to the ground-state energy E_0 of the form

$$N = T + V, \qquad\qquad (2.381)$$

$$E_0 \leq \frac{N}{D}, \quad T_{\lambda,\alpha} = \int_0^\infty dr\, r\big(\frac{d}{dr}e^{-\lambda r^\alpha}\big)^2, \quad V_{\lambda,\alpha} = \int_0^\infty dr \int_{-\pi}^{\pi} \frac{d\theta}{2\pi}\, V\, e^{-2\lambda r^\alpha},$$

$$D_{\lambda,\alpha} = \int dr\, r\, e^{-2\lambda r^\alpha}.$$

For $\lambda \to 0$, $V_{\lambda,\alpha}/D_{\lambda,\alpha}$ goes to $\int dx \int dy\, V$ uniformly for $0 < \alpha < 2$. $T_{\lambda,\alpha}/D_{\lambda,\alpha}$ becomes equal to $\alpha \int_0^\infty du\, u\, e^{-2u}$ by a change of variables and can be made arbitrarily small by taking small-enough α.

Before we discuss methods which allow the counting of bound states we may remind the readers of the somewhat surprising remark made in (2.3)–(2.5). Even if a potential goes to zero at infinity (but oscillates) a positive-energy bound state embedded in the continuum may exist.

This can occur only if the potential oscillates at infinity. We point out that there exist no positive-energy bound states if the potential is of bounded variation, or is absolutely integrable at infinity, in the sense that $\int_R^\infty |V(r)| dr < \infty$.

To show this we define the quantity $y = u'^2 + (E - V)u^2$ and derive from the Schrödinger equation that $y' = -u^2\, dV/dr$. We consider four cases:

(a) If $dV/dr < 0$ beyond some value of r which we denote by R, then y' is positive for all $r \geq R$. We can also assume $y(R)$ is positive because V goes to zero and $E > 0$. Hence $y(\infty) \neq 0$, which implies the non-occurrence of a bound state.

(b) If $dV/dr > 0$ beyond R we estimate

$$\frac{y'}{y} = -\frac{u^2\, dV/dr}{u^2(E - V) + u'^2} \geq -\frac{dV/dr}{E - \varepsilon} \qquad (2.382)$$

for some $\varepsilon > 0$, since $V \to 0$. We integrate (2.382) and obtain $\ln(y(\infty)/y(r)) > -(V(\infty) - V(r))/(E - \varepsilon)$, which again proves the assertion.

(c) If dV/dr changes sign we put $|dV/dr|$ in the above estimate, and this covers the case of a bounded variation.

(d) If we know only that $\int_R^\infty |V(r)| dr < \infty$, we define $Z = u'^2 + Eu^2$. Then $Z' = u^2 V$. Hence $|Z'/Z| < |V|/E$, and

$$\frac{Z(\infty)}{Z(R)} > \exp{-\frac{1}{E} \int_R^\infty |V(r)| dr} > 0,$$

which proves the impossibility of a positive-energy bound state.

If a Schrödinger operator with a local (scalar) potential has a ground state then (i) it is unique and (ii) the ground-state wave function will be positive everywhere.

We have a simple argument proving uniqueness. Assume that there are two ground-state wave functions to the same energy E_0:

$$(-\Delta + V)\psi_j = E_0\psi_j, \qquad j = 1, 2. \tag{2.383}$$

We may choose ψ_1 orthogonal to ψ_2: $\langle\psi_1|\psi_2\rangle = 0$. If ψ_1 is positive everywhere it follows that ψ_2 has to change sign somewhere. This leads to a contraction to the assertion that the ground-state wave function has to be positive. We still have to show the last assertion: the simplest situation is given in one dimension. Assume the ground-state wave function $\psi(x)$ changes sign. We take $|\psi_\varepsilon(x)|$ as a trial function where ε indicates a small smoothing procedure to round off the edge one obtains for $|\psi(x)|$. One loses a contribution to the kinetic energy proportional to $\delta\langle T\rangle \simeq -\varepsilon \cdot \ell$, where ℓ denotes the length of the region where we modified $|\psi(x)|$, while $\delta\langle V\rangle \simeq |\varepsilon|^2$. Putting both together we realize that we decrease the energy by taking $|\psi_\varepsilon(x)|$ as a trial function and that this is positive everywhere. A small variation of the argument applies to any dimension.

Birman–Schwinger bound

A certain bound on the number of bound states due to Birman and Schwinger is easy to obtain and is often helpful (see, for example, Ref. [67]), but has the wrong strong-coupling behaviour for large coupling. We transform the three-dimensional Schrödinger equation for bound states into an integral equation,

$$\psi(\vec{x}) = +\frac{1}{4\pi} \int d^3y \frac{e^{-\kappa|\vec{x}-\vec{y}|}}{|\vec{x} - \vec{y}|} |V(\vec{y})|\psi(\vec{y}), \tag{2.384}$$

where $-\kappa^2 = E$, and we assume that V is purely attractive. Otherwise V enters into (2.384) and $N(V) \leq N(V_-)$. We next define $\phi = \sqrt{V}\,\psi$ and transform (2.384) into an integral equation, which has a symmetric kernel,

$$\phi(\vec{x}) = \frac{1}{4\pi} \int d^3y \overbrace{\left(\frac{V^{1/2}(\vec{x})e^{-\kappa|\vec{x}-\vec{y}|}V^{1/2}(\vec{y})}{|\vec{x} - \vec{y}|}\right)}^{K(\vec{x},\vec{y})} \phi(\vec{y}). \tag{2.385}$$

Instead of considering the dependence of eigenfunctions as a function of energy, we can introduce a coupling constant λ into (2.385) and study the characteristic values λ_i:

$$\lambda_i\phi_{\lambda_i}(\vec{x}) = \int d^3y\, K(\vec{x}, \vec{y})\phi_{\lambda_i}(\vec{y}). \tag{2.386}$$

As a function of λ_i the eigenvalues move monotonously. If they cross $\lambda = 1$ they really represent eigenfunctions of the original problem. Instead of count the number of eigenvalues below zero energy of the original problem, it is equivalent to count the number of characteristic values λ_i greater than $\lambda = 1$. Therefore, the trace of K, being the sum of λ_is, will give an upper bound on N. Since that trace turns out to be infinite, we may bind $N < \sum_i \lambda_i^2$ by the Hilbert–Schmidt norm, which is easily calculated and yields

$$N \le \int \frac{d^3x d^3y}{(4\pi)^2} \frac{|V(\vec{x})||V(\vec{y})|}{|\vec{x} - \vec{y}|^2} . \tag{2.387}$$

It can be also shown that if (2.385) is finite, the total cross-section at positive energies is finite [131].

'Quasiclassical' estimates

The quasiclassical limit $\hbar \to 0$ for Schrödinger potential problems corresponds to the strong-coupling limit $\lambda \to \infty$, if we take λV instead of V as a potential. As already remarked (see (2.369)), we expect that all moments of energy levels will converge towards their classical phase-space integral. The main question now is whether inequalities of the type (2.369), with constants which might be equal to or greater than $C\ell_{\alpha,d}$, will hold. For an up-to-date review of this problem see the excellent article of Blanchard and Stubbe [132]. With the help of the Birman–Schwinger technique [127] the following result was obtained:

Theorem:
  Let $|V_-| \in L^{\alpha+d/2}(R^d)$ and $\alpha > 1/2$ for $d = 1$, $\alpha > 0$ for $d \ge 2$, or $\alpha \ge 0$ for $d \ge 3$. Let $e_j \le 0$ be the negative-energy bound states of $H = -\Delta + V(x)$. Then there exist finite constants $L_{\alpha,d}$, such that

$$\sum_j |e_j|^\alpha \le L_{\alpha,d} \int d^d x |V_-|^{\alpha+d/2} \tag{2.388}$$

holds.

Remarks:
  It is trivial to see that there cannot be a bound of the form (2.388) for $d = 1$ and $0 \le \alpha < 1/2$. Consider a potential like $v_\epsilon(z) = \theta(\epsilon - |z|)/2\epsilon$: for $\epsilon \to 0$, $v_\epsilon(z)$ goes to a δ-function potential with exactly one bound state, while the phase-space integral goes to zero. The same argument implies that the number of S-wave bound states n_0 for the three-dimensional Schrödinger problem cannot be bounded by the $||V||_{1/2}^{1/2}$ norm. A solution

has been found by Calogero and Cohn [133], who obtained a bound for monotonous potentials:

$$n_0 \leq \frac{2}{\pi} \int_0^\infty dr \, |V(r)|^{1/2} . \tag{2.389}$$

Only for certain values of α and d are the best possible answers to (2.388) known. Much effort was put into obtaining the best answer, for $d = 3$ and $\alpha = 1$ and $\alpha = 0$. In the latter case, one looks for a bound of the type

$$N_3 \leq C \int d^3x |V_-|^{3/2} . \tag{2.390}$$

The classical value is $C\ell_{0,3} = 1/6\pi^2 \simeq 0.0169$. The conjecture is that the Sobolev constant is the best possible answer: $K = 1/3\sqrt{3}\,(2/\pi)^2 \simeq 0.0780$. The best-known constant in an inequality of the form (2.390) is due to Lieb [134]: $0.1162 \simeq 1.49$ K. For the spherically symmetric problem we shall review our bound [135], which yields 0.1777 as a constant in (2.390). Li and Yau [136] obtained 0.5491 and Blanchard, Stubbe and Rezende [137] 0.1286.

Counting the number of bound states for spherically symmetric potentials

If the potential is rotationally symmetric (for reasons of simplicity let us treat the three-dimensional case) there is a simple way to count the number of bound states. The zero-energy wave function will divide the half-line into $n + 1$ parts through the positions where it vanishes. This node theorem holds only for $d = 1$ and half-line problems. We may use the inequality leading to (2.380) for each interval $[r_n, r_{n+1}]$ determined by these zeros, and obtain inequalities which, summed up, give

$$n_\ell \leq S_p \frac{\int_0^\infty dr(r^2V)^p}{(2\ell + 1)^{2p-1}} , \qquad p \geq 1, \quad S_p = \frac{(p-1)^{p-1}\Gamma(2p)}{p^p\Gamma^2(p)} . \tag{2.391}$$

Note that the Sobolev constant entering (2.377) is related to $S_{3/2}$: $4\pi S_{3/2} = s$. We wrote down the generalization taking into account angular momentum contributions and using Hölder's inequality for index p (see Appendix C). For $p = 1$ we obtain the old well-known bound of Bargmann. $p \to \infty$ gives a result of Courant–Hilbert.

In Ref. [135] a different way of counting the number of bound states is shown. Let us consider a rotationally symmetric problem and study the behaviour of eigenvalues as a function of ℓ:

$$\left(-\frac{d^2}{dr^2} + \frac{\ell(\ell + 1)}{r^2} - V(r) \right) u_{m,\ell}(r) = \varepsilon_{m,\ell} u_{m,\ell} . \tag{2.392}$$

The total number of bound states is given by $N = \sum_\ell (2\ell + 1)n_\ell$; for fixed m there is a maximal number of ℓ such that the m-th eigenvalue disappears in the continuum. Let us denote this value as ℓ_m with $[\ell_m]$ as the integer part of ℓ_m. Now we may count multiplicities: between $[\ell_m]$ and $[\ell_{m-1}] + 1$ we count $(2\ell + 1) \times 1$ bound states; between $[\ell_{m-1}]$ and $[\ell_{m-2}] + 1$ we count $(2\ell + 2) \times 2$ bound states and so on. Summing gives a formula for N where no approximations are involved (for $d = 3$),

$$N_3 = \sum_i ([\ell_i] + 1)^2 , \tag{2.393}$$

where ℓ_is are determined through the zero-energy Schrödinger equation

$$\left(-\frac{d^2}{dr^2} + \frac{\ell_i(\ell_i + 1)}{r^2} - V(r) \right) u_i = 0 . \tag{2.394}$$

As is well known, we may map problem (2.394) onto a problem defined on the full line $z \in (-\infty, \infty)$ by transforming from (r, u_i) to (z, ϕ_i) with $z = \ln r$ and $\phi_i(z) = (1/\sqrt{r}) u_i(r)$. We obtain

$$\left(-\frac{d^2}{dz^2} - v(z) \right) \phi_i(z) = -\left(\ell_i + \frac{d-2}{2} \right)^2 \phi_i(z) , \tag{2.395}$$

where we have written the general case for any $d \geq 2$. Since in $d = 3$ we may bind (2.393) by

$$N_3 \leq 4 \sum_i \left(\ell_i + \frac{1}{2} \right)^2 = 4 \sum_i |e_i| , \tag{2.396}$$

where the e_is are the eigenvalues of the one-dimensionnal Hamiltonian (2.395), the problem of obtaining bounds on the number of bound states in three dimensions is reduced to the problem of binding the first moment of energy levels in one dimension. But this can be done as described in Ref. [135]. Altogether, we obtain the result

$$N_3 \leq K \int d^3x \, |V_-|^{3/2} , \tag{2.397}$$

with $K \simeq 0.1777$, which is to be compared with the (probably best) answer of 0.0780.

As an aside we note that the analogue of (2.396) in $d = 4$ yields

$$N_4 \leq \sum |e_i|^{3/2} \leq \frac{3}{16} \int dz \, v^2(z) . \tag{2.398}$$

The second inequality in (2.398) comes from a sum rule [138] known to be optimal for reflectionless potentials. The bound in four dimensions is therefore optimal.

Next we mention the generalization of the above procedure to d dimensions. As a first step we relate N_d, the number of bound states in a potential problem in d dimensions, to a one-dimensional moment problem for energy levels.

For a rotationally invariant potential the reduced wave functions, analogous to (2.392), fulfil the equation

$$\left(-\frac{d^2}{dr^2} + \frac{(\ell + (d-1)/2)(\ell + (d-3)/2)}{r^2} - V(r)\right) u(r) = \epsilon\, u(r) . \quad (2.399)$$

The degeneracy of each level is determined by the number of harmonic polynomials of degree ℓ in d dimensions $D_\ell = H_\ell - H_{\ell-2}$, and the number of homogeneous polynomials of degree ℓ in d dimensions is given by

$$H_\ell = \binom{d + \ell - 1}{\ell} .$$

Now we again count multiplicities. If ℓ_m still denotes that angular momentum for which the m-th eigenvalue disappears in the continuum we obtain

$$N_d = \sum_i \sum_{\ell=0}^{[\ell_i]} (D_\ell - \ell) = \sum_i \frac{(d - 2 + [\ell_i])!(d - 1 + 2[\ell_i])}{(d-1)![\ell_i]!} . \quad (2.400)$$

As a second step we transform the problem into a purely one-dimensional one and obtain (2.395). Let us denote by $e_i = -(\ell_i + (d-2)/2)^2$ the eigenvalue of the one-dimensional problem corresponding to angular momentum ℓ_i. Next we bound (2.400) by

$$N_d \le K_d \sum_i |e_i|^{\frac{d-1}{2}}, \quad K_d = \sup_\ell \frac{(d - 2 + [\ell])!(d - 1 + 2[\ell])}{(d-1)![\ell]!([\ell] + (d-2)/2)^{d-1}} . \quad (2.401)$$

In order to proceed, we need bounds on moments of order 3/2, 2, 5/2, 3, etc. in one dimension. Aizenman and Lieb [139] proved that all moments of this type of order bigger than 3/2 are less than or equal to the appropriate classical constant times the appropriate moment of the potential:

$$\sum |e_i|^\alpha \le C\ell_{\alpha,1} \int dx |V(x)|^{\alpha+\frac{1}{2}}, \quad \alpha \ge 3/2 \quad (2.402)$$

$$C\ell_{\alpha,1} = \frac{\Gamma(\alpha + 1)}{\sqrt{4\pi}\,\Gamma(\alpha + 3/2)} .$$

$C\ell_{\alpha,d}$ has already been given in (2.369). The inequalities (2.401) and (2.402) together yield our results. For dimensions up to 20 they are summarized in Table 7.

The bounds thus obtained are compared to the one-particle bound S_d for $d \le 7$ and to the classical bound $C\ell_d$ for $d \ge 8$. The bounds are for all $d \ge 5$ larger than S_d and $C\ell_d$. Furthermore, for $d \ge 7$, the maximum of the r.h.s. of the expression determining K_d in Eq. (2.401) is reached for an ℓ value which is non-zero. Indeed, we even found calculable examples of potentials which show that for $d > 6$ the upper bound on N_d is necessarily

Table 7.   Illustration of inequality (2.401) and Eq. (2.407): Examples and bounds compared to the Sobelev and classical constant.

d	m_{opt}	Ex/S_d	$Ex/C\ell_d$	ℓ_{opt}	Bound$/S_d$	Bound$/C\ell_d$
3	1	1		0	2.28	
4	1	1		0	1	
5	1	1		0	1.08	
6	1	1		0	1.12	
7	3	1.16		1	1.23	
8	4		1.36	2		1.42
9	6		1.29	4		1.33
10	8		1.24	5		1.27
11	10		1.21	7		1.23
12	13		1.18	10		1.21
13	16		1.16	13		1.18
14	20		1.15	16		1.16
15	24		1.13	19		1.14
16	28		1.12	23		1.14
17	32		1.11	27		1.13
18	37		1.11	32		1.12
19	42		1.10	37		1.10
20	47		1.09	42		1.10

above the classical bound, but approaches it for $d \to \infty$. This disproves a conjecture of Lieb and Thirring[127]: the belief was that there exists a critical value α in (2.388) called $\alpha_{c,d}$, such that

$$L_{\alpha,d} = L_{\alpha,d}^1 \text{ for } \alpha \leq \alpha_{c,d} \text{ and } L_{\alpha,d} = L_{\alpha,d}^c \text{ for } \alpha \geq \alpha_{c,d}, \qquad (2.403)$$

where $L_{\alpha,d}^1$ denotes the optimal constant in (2.388) for $N = 1$ and $L_{\alpha,d}^c$ denotes the classical constant. It has been proven that $\alpha_{c,1} = 3/2$; it was conjectured by numerical experiments that $\alpha_{c,2} = 1.165$ and $\alpha_{c,3} = 0.8627$. We shall describe the examples which disprove the conjecture for $\alpha = 0$ and $d \geq 7$ in the next subsection.

A variational approach

We have also tried to obtain an optimal result. In principle, we intend to evaluate a supremum of a certain functional, overall potentials which support a number of bound states. If we fix this number N_d we therefore would like to obtain the $\inf_V \int d^d x\, V(x)^{d/2}$. It is clear that there has to exist at least one zero-energy bound state. Otherwise we could decrease the coupling constant, changing the above-mentioned functional without changing the number of bound states. Let us assume that this zero-energy bound state occurs for angular momentum L. The variational principle which we arrive at reads

$$\delta \int d^d x \, V(x)^{d/2} \geq 0, \qquad \delta E_L = \int d^d x \, \psi_L^2(x) \delta V \leq 0, \qquad (2.404)$$

where E_L denotes the eigenvalue of the particular bound state. Variation of V has to be such that this bound-state energy remains zero or becomes negative.

We introduce a Lagrange multiplier and derive the proportionality $|V|^{(d-2)/2} = \text{const}|\psi|^2$. This implies a scale (and conformal) invariant non-linear field equation for the zero-energy wave function ψ_L:

$$-\Delta \psi_L - |\psi_L|^{4/(d-2)} \psi_L = 0. \qquad (2.405)$$

For $d = 3$ all solutions of (2.405) are known. $L = 0$ gives the optimal answer and the Sobolev constant results in the bound for N_d. Unfortunately, there may also be other solutions to a variational principle with more zero-energy eigenvalues with angular momentum L_j. The resulting non-linear equation, replacing (2.405), becomes

$$-\Delta \psi_{L_i} - \left(\sum_j |\psi_{L_j}|^2 \right)^{2/(d-2)} \psi_{L_i} = 0. \qquad (2.406)$$

We obtained a number of interesting solutions to (2.406) by projecting stereographically onto a d-dimensional sphere and using the completeness property of spherical harmonics on that sphere. With the help of these solutions we have been able to gain insight into the questions asked, but the proof that we have obtained all solutions relevant to the problem has not been given.

As we have already mentioned, there may occur a number of zero-energy states for certain angular momenta. We can now sum up the multiplicities following (2.400) and obtain for m zero-energy bound states

$$\frac{N_d}{I_d \, C\ell_d} = \frac{2^{d-1}(d+m-2)!(d+2m-2)}{[(d+2m-2)(d+2m-4)]^{\frac{d}{3}}(m-1)!}, \qquad (2.407)$$

where $I_d = \int_{-\infty}^{\infty} dz \, |v(z)|^{d/2}$. The results of the optimization in m are also given in Table 7. For $3 \leq d \leq 7$ we have to compare these with the Sobolev constants S_d. For $d \geq 7$ our examples violate the Sobolev bound which would result if the nodal theorem were true. We believe that the numbers obtained from (2.407) after optimization of m are the best possible ones. So the new conjecture is that up to $d = 6$ the Sobolev constants will be the best possible. For larger d the best answer will be given by the optimizing the m in (2.407) and for $d \to \infty$ the classical value will be reached.

It should be noted that the optimal answers to the mentioned bounds in $d = 3$ are still missing. In $d = 4$ the system (2.406) has been shown to be completely integrable [140]. This field theory is related to the conformal invariant instanton equation.

Number of bound states in oscillating potentials of the Chadan class

We have already mentioned in Section 2.1 that there is a class of oscillating potentials for which the Schrödinger equation is perfectly well behaved. This class was discovered by Chadan [66] some years ago, and, from the point of view of principles it is remarkable, because what Chadan requires are integrability properties of the absolute value of the primitive of the potential rather than of the absolute value of the potential itself. Here, we shall restrict ourselves to potentials oscillating near the origin and going to zero sufficiently rapidly at infinity (though one can also consider the case of potentials oscillating at infinity [141]). In this case one defines

$$W(r) = \int_r^\infty V(r')dr' \qquad (2.408)$$

and the Chadan class is defined by

$$\begin{cases} \lim_{r\to 0} r\, W(r) = 0 \\ \int_r^R |W(r')|dr' < \infty \text{ for } r \to 0 \,. \end{cases} \qquad (2.409)$$

We see that ordinary potentials such that $V \sim r^\alpha$ for $r \to 0$ with $\alpha > -2$ will satisfy these requirements. If, in addition, we require $|W(r)| < r^{-\epsilon}$ for $r \to \infty$, we have all desired properties and, in particular, all bound states have negative energy.

We shall show that it is very easy to get bounds on the number of bound states in potentials of the Chadan class [142]. Suppose the zero-energy solution of the Schrödinger equation which is regular at the origin has n nodes — i.e., we have n bound states. Consider the interval $r_k < r < r_{k+1}$ where r_k and r_{k+1} are successive nodes. Then, integrating the Schrödinger equation from r_k to r_{k+1}, we get

$$0 = \int_{r_k}^{r_{k+1}} u'^2(r)dr + \int_{r_k}^{r_{k+1}} \left(\frac{\ell(\ell+1)}{r^2} + V(r) \right) u^2(r)dr \,.$$

We then integrate by parts using the definition of W, noting that the integrated terms vanish:

$$0 = \int_{r_k}^{r_{k+1}} \left[u'^2(r) + \frac{\ell(\ell+1)}{r^2}u^2(r) \right] dr + 2 \int_{r_k}^{r_{k+1}} W(r)u(r)u'(r)dr \qquad (2.410)$$

and use the inequality

$$|2W(r)u(r)u'(r)| < \frac{1}{2}|u'(r)|^2 + 2|W(r)u(r)|^2 \,.$$

Hence we have

$$0 > \int_{r_k}^{r_{k+1}} \left[\frac{1}{2}|u'(r)|^2 + \frac{\ell(\ell+1)}{r^2}u^2(r) - 2|W(r)|^2 u^2(r) \right] dr$$

or

$$0 > \int_{r_k}^{r_{k+1}} \left[|u'(r)|^2 + \frac{L(L+1)}{r^2}u^2(r) - 4|W(r)|^2 u^2(r) \right] dr , \qquad (2.411)$$

with $L(L+1) = 2\ell(\ell+1)$. Inequality (2.411), seen as a variational inequality, means that inside the potential $-4|W(r)|^2$, with angular momentum L, and with Dirichlet boundary conditions at r_k and r_{k+1}, there is one negative-energy bound state. Hence, any one of the conditions (2.391) applies with the necessary substitutions of ℓ by L and $|V|$ by $4W^2$. For instance, we have

$$\frac{1}{2L+1} \int_{r_k}^{r_{k+1}} 4r W^2(r) dr > 1 ,$$

and reexpressing L and adding up all the intervals we get

$$n_\ell < \frac{1}{2\sqrt{2(\ell+1/2)^2 - 1/4}} \int_0^\infty 4r|W(r)|^2 dr ,$$

and similarly

$$n_\ell < \frac{16}{3\sqrt{3}\pi} \frac{\int 8r^2 |W(r)|^3 dr}{8(\ell+1/2)^2 - 1} .$$

In fact there is some sort of self-consistency: if V is purely attractive — i.e., negative — $W(r)$ is negative, monotonous increasing. Hence, the Calogero bound

$$n_0 < \frac{2}{\pi} \int_0^\infty \sqrt{|U(r)|} dr ,$$

which is valid for a potential $U(r)$ negative, monotonous increasing, applies. So we get

$$n_0 < \frac{4}{\pi} \int_0^\infty |W(r)| dr = \frac{4}{\pi} \int_0^\infty r|V(r)| dr ,$$

which is, except for a factor larger than unity, the Bargmann bound.

These bounds are not exactly optimal. Bounds with optimal constant can be found in original references.

3

Miscellaneous results on the three-body and N-body problem

The successful application of the Schrödinger equation to quark–antiquark systems implies, unavoidably, that one should also apply it to baryons that are systems of three quarks, and which, because of 'colour', are not constrained to have an antisymmetric wave function in spin and space. In fact, historically it was rather the contrary, in the sense that the quark model was first applied to baryons [13]. We shall not describe here the details of the model calculations, but shall rather describe some general properties coming from the belief that forces between quarks are flavour-independent, as well as from the postulate that the Schrödinger equation holds.

We have said already that if a Hamiltonian is of the form $H = A + \lambda B$ its ground-state energy is concave in λ — i.e., $d^2 E(\lambda)/d\lambda^2 \leq 0$. A very simple application to the two-body problem [101], [143] is that the binding energy of a system $Q_1 \bar{Q}_2$ whose Hamiltonian is

$$-\frac{1}{2} \left[m_{Q_1}^{-1} + m_{Q_2}^{-1} \right]^{-1} \frac{d^2}{dr^2} + V(r)$$

is concave in $[m_{Q_1}^{-1} + m_{Q_2}^{-1}]^{-1}$. This means that

$$E_{Q_1 \bar{Q}_2} \geq \frac{E_{Q_1 \bar{Q}_1} + E_{Q_2 \bar{Q}_2}}{2} \tag{3.1}$$

and, adding the constituent masses,

$$M_{Q_1 \bar{Q}_2} \geq \frac{1}{2} (M_{Q_1 \bar{Q}_1} + M_{Q_2 \bar{Q}_2}) . \tag{3.2}$$

This inequality works very well. For instance, assuming that spin-dependent forces do not cause complications, we have

$$M_{D^{\cdot}} = 2.12 \text{ GeV} \geq \frac{1}{2} (M_{J/\psi} + M_\phi) = 2.06 \text{ GeV} .$$

The same concavity property applies for three-body systems — for

136

instance, baryons made of three quarks. In particular, $E(Q_1, q_2, q_3)$, the binding energy of a system of one heavy quark and two light quarks is concave in $1/m_{Q_1}$, since the Hamiltonian is

$$-\frac{1}{2m_{Q_1}} \Delta_1 - \frac{1}{2m_{q_2}} \Delta_2 - \frac{1}{2m_{q_3}} \Delta_3 + \sum V_{ij} . \tag{3.3}$$

If one has a model giving constituent quark masses, one can in this way get an upper bound for the mass of a baryon containing a b-quark when the masses of the similar baryons containing c- and s-quarks are known. For instance:

$$M_{\Lambda_b} < m_b + \frac{(M_{\Lambda_c} - m_c)\,(1/m_c - 1/m_b)}{1/m_s - 1/m_c} - \frac{(m_\Lambda - m_s)\,(1/m_s - 1/m_b)}{1/m_s - 1/m_c} ,$$

and with $m_s = 0.518$ GeV/c^2, $m_c = 1.8$ GeV/c^2, $m_b = 5.174$ GeV/c^2 [34], this gives a reasonable value, Ref. [144],

$$M_{\Lambda_b} < 5.629 \text{ GeV}/c^2$$

and similarly

$$M_{\Sigma_b} < 5.826 \text{ GeV}/c^2 .$$

At present, experiments give a mass of 5.625 GeV for the Λ_b [145].

It is very tempting to go further in these concavity properties, and change all masses in

$$H = -\frac{1}{2m_{Q_1}} \Delta_1 - \frac{1}{2m_{Q_2}} \Delta_2 - \frac{1}{2m_{Q_3}} \Delta_3 + V_{12} + V_{23} + V_{13} . \tag{3.4}$$

Richard and Taxil [44] have observed that the members of the SU_3 decuplet

$$\left. \begin{array}{l} (q\,q\,q)\,\Delta \\ (s\,q\,q)\,\Sigma^* \\ (s\,s\,q)\,\Xi^* \\ (s\,s\,s)\,\Omega^- \end{array} \right\} \tag{3.5}$$

have an energy and a mass which is a concave function of the strangeness of the baryon for a 'reasonable' potential (with no spin-dependent forces) — i.e.,

$$E(m_1, m_2, m_2) > \frac{1}{2}[E(m_2, m_2, m_2) + E(m_1, m_1, m_2)] . \tag{3.6}$$

But what is 'reasonable'? Some 'general' incorrect proofs were proposed, but in the end Lieb [45] gave a proof for a restricted class of potentials. This class is such that

$$\exp - \beta V(r)$$

should have a positive three-dimensional Fourier transform; two sufficient conditions for this to hold were proposed by Lieb.

(a) V, considered as a function of r^2, should have successive derivatives with alternate signs [45] ($dV/dr^2 > 0$ etc...);

(b)

$$V' > 0, \quad V'' < 0, \quad V''' > 0, \tag{3.7}$$

a condition previously proposed by Askey [146].

In fact, condition (b), as shown by one of us (A.M.) [147] can be slightly weakened:

(c)

$$V' > 0, \quad V'' < 0, \quad rV''' - V'' \ge 0. \tag{3.8}$$

This latter condition is optimal in a certain sense that we shall explain. Let F be such that $F > 0$, $F' < 0$, $F'' > 0$. Then, by successive integrations by parts, the three-dimensional Fourier transform of F, $\tilde{F}(q)$ is found to be proportional to

$$\int_0^\infty dr \left(\frac{F''(r)}{r^2} - \frac{F'''(r)}{r} \right) \frac{[qr(1 + \cos qr) - 2 \sin qr]^2}{1 + \cos qr}. \tag{3.9}$$

If the support of $F'' - rF'''$ is restricted to points where the square bracket vanishes, the Fourier transform vanishes.

Naturally one can ask oneself if for some potentials not satisfying these conditions the concavity property could be violated. As shown by Lieb in the extreme case of one particle of infinite mass and a square well this is the case. It is also the case as shown <u>analytically</u> by Richard, Taxil and one of the present authors (A.M.) [44] for a potential $V = r^5$, and <u>numerically</u> for $V = r^{2.6}$.

These potentials, however, are not 'physical', since they are not concave in r [46]. Yet the weakest condition found involves the third derivative of the potential. It is not known if this is due to a deficiency of mathematical physicists or if it is deeply needed. We tend to favour the former assumption. The 'equal spacing' of the decuplet results from a compensation between the concave central energy and the spin-dependent forces.

It is tempting to go further than (3.6), following the suggestion of Richard, and try to prove the inequality

$$E(m_1, m_2, m_3) > \frac{1}{3}[E(m_1, m_1, m_1) + E(m_2, m_2, m_2) + E(m_3, m_3, m_3)]. \tag{3.10}$$

This turns out to be very easy to prove [148], using the method of Lieb [45], who was able to prove an inequality more general than (3.6):

$$E(m_1, m_2, m_3) > \frac{1}{2}[E(m_1, m_1, m_3) + E(m_2, m_2, m_3)] , \qquad (3.11)$$

under the same conditions — (3.8) for instance. It is also possible to find counterexamples with sufficiently rapidly rising potentials — for instance, $V(r) = r^{4.9}$.

Other very interesting inequalities on three-body systems can be obtained if one accepts a not completly justified assumption: in a baryon the quark–quark potential is related to the corresponding quark–antiquark potential by

$$V_{Q_1 Q_2} = \frac{1}{2}V_{Q_1 \bar{Q}_2} . \qquad (3.12)$$

The basis for this 'belief' is that it holds for the one-gluon exchange potential. Also, if in a baryon two quarks are close to one another, they form a diquark, which is in a $\bar{3}$ representation, since it must combine with the representation 3 of the remaining quark. In all respects it behaves like an antiquark and its potential interaction with the quark is $V_{Q\bar{Q}}$. Dividing by two as there are two quarks, we get rule (3.12). From a phenomenological point of view, this rule has led to a beautiful prediction of the mass of the Ω^- particle by Richard [41] based entirely on the fit [34] of quarkonium, which is $M_{\Omega^-} = 1666$ MeV/c^2 (while experiment gives 1672 MeV/c^2).

Once the rule is accepted, it has beautiful consequences. The three-body Hamiltonian

$$H_{123} = -\sum \frac{1}{2m_{Q_i}}\Delta_i + \sum V_{ij} \qquad (3.13)$$

can be written as

$$H_{123} = \frac{1}{2}\sum_{i,j} H_{ij} , \qquad (3.14)$$

with

$$H_{ij} = -\frac{1}{2m_i}\Delta_i - \frac{1}{2m_j}\Delta_j + 2V_{ij} . \qquad (3.15)$$

H_{ij} is, according to rule (3.12), a quark–antiquark Hamiltonian. We have

$$\inf H_{123} > \frac{1}{2}[\inf H_{12} + \inf H_{23} + \inf H_{13}] .$$

Now including the constituent masses we get [40]

$$E(Q_1, Q_2, Q_3) > \frac{1}{2}[E(Q_1, \bar{Q}_2) + E(Q_1, \bar{Q}_3) + E(Q_2, \bar{Q}_3)]$$

and

$$M(Q_1, Q_2, Q_3) > \frac{1}{2}[M(Q_1, \bar{Q}_2) + M(Q_1, \bar{Q}_3) + M(Q_2, \bar{Q}_3)] . \tag{3.16}$$

Strictly speaking, this relation holds for particles without spin. But it also holds for parallel spins. In this way we get

$$M_\Delta > 3/2 M_\rho ,$$

$$M_{\Omega^-} > 3/2 M_\phi .$$

In Ref. [149] Richard and one of the authors (A.M.) have been able to incorporate spin as perturbation. One gets:

Spin 3/2 states

$$\Omega^- = 1.672 > 3/2 \, \phi = 1.530$$
$$\Delta = 1.232 > 3/2\rho = 1.155$$
$$\Sigma^* = 1.385 > K^* + 1/2\rho = 1.275$$
$$\Xi^* = 1.532 > \rho + 1/2\phi = 1.280$$
$$\Sigma_b^* > B^* + 1/2\rho = 5.710$$

Spin 1/2, Σ-like states

$$N = 0.938 > 3/4\rho + 3/4\pi = 0.681$$
$$\Sigma = 1.190 > 1/2\rho + 3/4K + 1/4K^* = 0.979$$
$$\Xi = 1.319 > 1/2\phi + 3/4K + 1/4K^* = 1.104$$
$$\Sigma_c = 2.450 > 1/2\rho + 3/4D + 1/4D^* = 2.288$$
$$\Sigma_b > 1/2\rho + 3/4B + 1/4B^* = 5.670$$

Spin 1/2, Λ-like states

$$\Lambda = 1.116 > 1/2\pi + 3/4K^* + 1/4K = 0.863$$
$$\Lambda_c = 2.286 > 1/2\pi + 3/4D^* + 1/4D = 2.042$$
$$\Lambda_b > 1/2\pi + 3/4B^* + 1/4B = 5.379 .$$

$$\tag{3.17}$$

We see that the deviations between the l.h.s. and the r.h.s. of Eqs. (3.17) oscillate between 100 and 250 MeV. The question is whether one can do better than this. That is to say: Can we improve the lower bound given by (3.16)? The answer is yes.

The method we shall use is, in fact, very general and can be generalized to the N-body case without the need to restrict oneself to the three-body case. However, in its simplest form it works for equal-mass particles. The unequal-mass case has been worked out completely for the three-body case only.

We start with N equal-mass particles. We use the notation $E_N(M, V)$ to designate the energy of N particles of mass M interacting with a potential

V. In this language, the previous inequality for equal masses is

$$E_3\left(M, \frac{V}{2}\right) > \frac{3}{2}E_2(M, V).\tag{3.18}$$

Notice that

$$E_N(M, V) \equiv \frac{1}{\lambda}E_N\left(\frac{M}{\lambda}, \lambda V\right).\tag{3.19}$$

The new inequality [9, 150] is based on a very simple identity on the kinetic energy:

$$\sum_{i>j}(p_i - p_j)^2 + \left(\sum p_i\right)^2 = N\sum p_i^2.\tag{3.20}$$

Thus, disregarding the centre-of-mass energy $(\sum p_i)^2$ we can write $H_N(M, V)$, the Hamiltonian of N particles of mass M with pairwise potential V, as

$$\left.\begin{array}{l} H_N(M, V) = \sum h_{ij} \\[2mm] \text{with}\quad h_{ij} = \dfrac{(p_i - p_j)^2}{2MN} + V(r_{ij}) \end{array}\right\},\tag{3.21}$$

but $p_i - p_j = 2\pi_{ij}$ is TWO times the conjugate momentum to r_{ij}:

$$h_{ij} = \frac{2(\pi_{ij})^2}{MN} + V(r_{ij}).\tag{3.22}$$

Therefore h_{ij} is a Hamiltonian corresponding to a reduced mass $MN/4$ — i.e., two particles of mass $MN/2$. Hence we get

$$E_N(M, V) > \frac{N(N-1)}{2}E_2\left(\frac{MN}{2}, V\right).\tag{3.23}$$

It is easy to see that (3.23) becomes an equality in the case of harmonic oscillator forces [151]. For $N = 3$ this gives

$$E_3(M, V) > 3E_2\left(\frac{3}{2}M, V\right),$$

or, using scaling properties,

$$E_3\left(M, \frac{V}{2}\right) > \frac{3}{2}E_2\left(\frac{3}{4}M, V\right).\tag{3.24}$$

If we compare this with (3.18) we see that M has been replaced by $(3/4)M$. The mass is reduced and hence the lower bound is higher.

The gain obtained with this inequality can be appreciated both numerically and analytically. Numerically, one can calculate extremely accurate variational upper bounds for power potentials and then compare with the old and new lower bounds. This is done in Table 8.

Table 8.   Ground-state energy E_3 of $H_3 = \Sigma(1/2)p_i^2 + (1/2)\epsilon(\beta)\Sigma r_{ij}^{\beta}$ compared to the naive limit $(3/2)E_2(1;r^{\beta})$, the improved lower limit $(3/2)E_2((3/4);r^{\beta})$ and a simple variational approximation $\tilde{E}_3$ obtained with a Gaussian wave function.

β	$\frac{3}{2}E_2(1)$	$\frac{3}{2}E_2(\frac{3}{4})$	E_3	$\tilde{E}_3$
-1	-0.37500	-0.28125	-0.26675	-0.23873
-0.5	-0.65759	-0.59746	-0.59173	-0.57964
0.1	1.85359	1.87916	1.88019	1.88278
0.5	2.75009	2.91296	2.91654	2.92590
1	3.50716	3.86013	3.86309	3.87114
2	4.5	5.19615	5.19615	5.19615
3	5.17584	6.15098	6.15591	6.17147

Analytically, there are ways to evaluate the difference $E_2(3/4\,M, V) - E_2(M, V)$ by using the inequalities on the kinetic energy derived in Section 2.4

$$\langle T \rangle > \frac{3}{4}(E_p - E_s) , \qquad (3.25)$$

where E_s denotes the $\ell = 0$ ground-state energy of a two-body system, and E_p the energy of the lowest $\ell = 1$ state, and the property, following from the Feynman–Hellmann theorem:

$$\frac{dE}{d\mu}(M, V) = -\frac{\langle T \rangle}{\mu} . \qquad (3.26)$$

By including some further concavity properties of $\mu \langle T \rangle$ one can integrate (3.26) and get a lower bound on the difference of binding energy between a two-body system of two masses $(3/4)M$ and two masses M. This gives

$$E_3 \left(M, \frac{V}{2} \right) > \frac{3}{2} \left[E_2(M, V, \ell = 0) + \frac{3}{16} \left[E_2(M, V, \ell = 1) - E_2(M, V, \ell = 0) \right] \right]. \qquad (3.27)$$

A rather spectacular application can be made to the Ω^- systems: the previous inequality gave

$$M_{\Omega^-} > 1530 \text{ GeV} ;$$

the new inequality gives

$$M_{\Omega^-} > 1659 \text{ MeV}$$

to be compared with the experimental value:

$$M_{\Omega^-} = 1672 \text{ MeV} .$$

Table 9. Comparison of the various lower bounds for $V = r$, and masses $(1, 1, M)$. We show the ratio $E^{(3)}/E^{(2)}$. The variational bound corresponds to a hyperspherical calculation up to grand orbital momentum $L = 8$.

M	Variational	Naïve	Optimized
0.05	5.7178	5.3795	5.7141
0.1	4.8724	4.5303	4.8693
0.2	4.2250	3.8845	4.2220
0.5	3.6147	3.2894	3.6120
1	3.3045	3.0000	3.3019
2	3.0976	2.8171	3.0953
5	2.9408	2.6869	2.9386
10	2.8795	2.6386	2.8773
20	2.8466	2.6134	2.8444

Table 10. Comparison of the various lower bounds for $V = -r^{-1}$, and masses $(1, 1, M)$. We show the ratio $E^{(3)}/E^{(2)}$, so that lower bounds become upper limits and vice versa.

M	Variational	Naïve	Optimized
0.05	0.67500	1.1905	0.69332
0.1	0.84693	1.3636	0.86868
0.2	1.1183	1.6667	1.1609
0.5	1.6484	2.3333	1.7341
1	2.1340	3.0000	2.2500
2	2.6045	3.6667	2.7427
5	3.0787	4.3333	3.2341
10	3.2990	4.6364	3.4620
20	3.4268	4.8095	3.5944

All this can be generalized to three particles of unequal masses [151]. To do this one has to exploit the freedom to add arbitrarily to the Hamiltonian terms of the form $(P \cdot \Sigma \alpha_i\, p_i)$, where

$$P = \sum_{i=1}^{3} p_i \,.$$

The results of numerical experiments for power potentials are again spectacular, as shown in Tables 9 and 10.

Now let us return to the general N-body system. A particularly spectacular application is that of N particles interacting with a gravitational potential [9]:

$$H = \sum_i^N \frac{p_i^2}{2m} - \sum_{i>j}^N \frac{x}{r_{ij}},$$

(3.28)

with $x = Gm^2$, G being the Newton constant. An upper bound on the energy of this N-particle system can be obtained either by taking a hydrogen-like trial function using the variable $\sqrt{\Sigma |r_{ij}|^2}$, which is good for small N, or a Hartree wave function $\Pi_i(f(r_i))$, which is good for large N, when centre-of-mass effects are negligible.

This gives

$$E_3 < -1.0375 \, G^2 m^5$$

$$E_N < -0.0542 \, N(N-1)^2 G^2 m^5,$$

with

$$f(r) = \exp{-\sqrt{ar^2 + b}} \; ;$$

the new lower bound gives

$$E_3 > -1.125 \, G^2 m^5$$

$$E_N > -0.0625 \, N^2(N-1) \, G^2 m^5 \; .$$

We see that the upper and lower bounds agree within 15%, which is rather remarkable.

To finish this chapter, we would like to present a very interesting and intriguing development due to Gonzalez-Garcia [152]. He combines inequalities (3.23) and (3.24) with a $1/d$ expansion of the energies, where d is the dimension of the space.

The idea of a $1/d$ expansion is that if we take N particles in d dimensions $d \gg N$, the system, provided the forces are conveniently scaled (which is easy for power potentials), is 'frozen', with fixed distances between the particles. Deviations from these equilibriuim positions can be described by harmonic motions, giving rise to $1/d$ corrections. There is, so far, no rigorous control on these $1/d$ expansions.

The contribution of Gonzalez-Garcia is that the ratio

$$R = \frac{E_N(M,V)}{E_2(NM/2,V)N(N-1)/2}$$

(3.29)

approaches unity not only when V becomes a harmonic oscillator potential, but also when $d \to \infty$. The idea, then, is to expand R in powers of $1/d$ instead of expanding the individual energies. If this is done to the lowest

Table 11. Expansion of R of Eq. (3.29) to lowest
order in $1/d$ compared to the numerical values.

β	$R_{1/d}$	$R_{\text{numerical}}$ (Table 8)	Relative error ‰
-1	0.946	0.948	2
-0.5	0.9898	0.9904	0.6
0.5	1.0016	1.0012	0.3
1	1.0011	1.0008	0.3
2	1	1	0
3	1.0013	1.0008	0.4

non-trivial order in $1/d$, it gives the results presented in Table 11 for a three-body system with equal masses, with two-body potential $\epsilon(\beta)r^\beta$ (see Table 8).

These results are really impressive, even though we have no idea of the convergence (or asymptotic character?) of the $1/d$ expansion.

Similarly, for a system of particles in gravitational interaction, Gonzalez-Garcia obtains

$$\lim_{N\to\infty} \frac{E_N}{N^3 G^2 m} = -0.0539 \,,$$

to be compared with our upper bound, which is -0.0542.

It would certainly be desirable to understand the reasons for the success of the $1/d$ approach. This is an interesting, but difficult problem.

We have presented only a few facets of the three- and N-body problems. We recommend to the reader the excellent review of Richard [153] on baryons as three-quark systems.

Appendix A
Supersymmetric quantum mechanics

Here we introduce the appropriate algebraic scheme and indicate the connection to factorization of Schrödinger operators used in our discussion of level ordering.

We shall start with the simplest system known to physics, namely the harmonic oscillator, which we may call a bosonic one with Hamiltonian

$$H_B = -\frac{\omega}{2}\frac{d^2}{dx^2} + \frac{\omega}{2}x^2 . \tag{A.1}$$

For reasons of simplicity we have put $m = 1/\omega$. We rewrite (A.1) in 'factorized' form

$$H_B = \omega\left(a^\dagger a + \frac{1}{2}\right) , \quad a = \left(\frac{d}{dx} + x\right)\frac{1}{\sqrt{2}} , \quad a^\dagger = \left(-\frac{d}{dx} + x\right)\frac{1}{\sqrt{2}} , \tag{A.2}$$

where $a^\dagger$ and a denote creation and annihilation operators for quanta of the oscillator. Their commutation relations,

$$[a, a^\dagger] = 1 , \qquad [H_B, a^\dagger] = \omega a^\dagger , \tag{A.3}$$

show the 'bosonic' nature of H_B. The spectrum and associated eigenfunctions are given by

$$E_n = \omega\left(n + \frac{1}{2}\right) , \quad n \in \mathbf{N}_0 , \quad a|0\rangle = 0 , \quad \frac{(a^\dagger)^n}{\sqrt{n!}}|n\rangle , \tag{A.4}$$

where $|0\rangle$ denotes the ground-state vector and $\omega/2$ is the zero-point energy.

Before we realize the simplest form of a supersymmetry within potential models, we 'double' the system treated before by adding a second bosonic oscillator:

$$h = \frac{1}{2}(p_1^2 + \omega_1^2 q_1^2) + \frac{1}{2}(p_2^2 + \omega_2^2 q_2^2) . \tag{A.5}$$

This two-dimensional oscillator obviously has an additional symmetry if $\omega_1 = \omega_2$. Then h is invariant under rotations within the plane, $q_1, q_2,$

generated by the angular momentum operator

$$\ell = q_1 p_2 - q_2 p_1, \qquad\qquad [\ell, h] = 0. \qquad (A.6)$$

We may replace the two real coordinates q_1, q_2 by one complex one $q = q_1 + iq_2$. The above-mentioned rotation becomes the phase transformation $q \rightarrow e^{i\varphi} q$. The invariance of h implies the existence of a conserved charge. In the above case the angular momentum is the conserved quantity. ℓ acts in the space of the two bosonic degrees of freedom and it is easy to exponentiate ℓ to obtain the unitary group implementing the transformation.

In contrast to the doubled bosonic oscillator we add now a 'fermionic' degree of freedom to H_B of (A.1) and obtain the supersymmetric oscillator

$$H = \omega(a^\dagger a + c^\dagger c), \qquad (A.7)$$

which acts on pairs of square integrable functions. The $c^\dagger$ and c denote fermionic creation and annhilation operators and obey anticommutation relations

$$\{c, c^\dagger\} = 1, \qquad\qquad c^{\dagger 2} = c^2 = 0. \qquad (A.8)$$

If we started with the Hamiltonian $\omega_1(a^\dagger a + \frac{1}{2}) + \omega_2(c^\dagger c - \frac{1}{2})$ instead of (A.7), no symmetry would be built in. We note that the zero-point energy of the fermionic oscillator is negative. For equal frequencies $\omega_1 = \omega_2 = \omega$, the zero-point energies of the bosonic and the fermionic oscillator cancel.

The algebra (A.8) can be well represented by σ-matrices, since $\{\sigma^-, \sigma^+\} = 1$, $\sigma^{+2} = \sigma^{-2} = 0$. This representation is used in (A.7) and H can therefore be considered to be a 2×2 matrix operator. For more degrees of freedom we have to introduce a Klein–Jordan–Wigner transformation in order to represent the algebra of operators, which obey anticommutation relations, by the matrices of the form $\mathbf{1} \otimes \mathbf{1} \otimes \ldots \otimes \sigma^j \otimes \mathbf{1} \otimes \ldots \otimes \mathbf{1}$.

We may rewrite (A.7) as

$$H = \omega\{a^\dagger a(1 - c^\dagger c) + (a^\dagger a + 1)c^\dagger c\} = \omega(a^\dagger a c c^\dagger + a a^\dagger c^\dagger c) = \{Q, Q^\dagger\}, \quad (A.9)$$

where we have introduced operators Q and $Q^\dagger$ through

$$Q = \sqrt{\omega}\, a c^\dagger, \qquad\qquad Q^\dagger = \sqrt{\omega}\, a^\dagger c, \qquad (A.10)$$

which turn out to act like charges. They are called supercharges and can be written for our simple case as

$$Q = \sqrt{\omega} \begin{pmatrix} 0 & a \\ 0 & 0 \end{pmatrix}, \qquad Q^\dagger = \sqrt{\omega} \begin{pmatrix} 0 & 0 \\ a^\dagger & 0 \end{pmatrix},$$

$$\{Q, Q^\dagger\} = \omega \begin{pmatrix} a a^\dagger & 0 \\ 0 & a^\dagger a \end{pmatrix} = H. \qquad (A.11)$$

The supercharges determine a special case of a square root of the Hamiltonian H. A simple calculation shows that they commute with the Hamiltonian:

$$[Q, H] = \omega^{3/2}[ac^\dagger, a^\dagger a + c^\dagger c] = \omega^{3/2}(c^\dagger a - ac^\dagger) = 0. \qquad (A.12)$$

They generate a symmetry of the Hamiltonian which is called a 'supersymmetry'. We remark at this stage that H consists of the direct sum of two operators $a^\dagger a$ and $aa^\dagger$ which are 'essentially' isospectral. This notion means that their spectra coincide except for zero modes

$$\text{spectr}\,(a^\dagger a) \setminus \{0\} = \text{spectr}\,(aa^\dagger) \setminus \{0\}. \qquad (A.13)$$

For the pure-point spectrum, (A.13) is obtained from the simple observation that $a^\dagger a \psi = E\psi \Rightarrow aa^\dagger a\psi = Ea\psi$, and $a\psi$ is therefore an eigenfunction for $aa^\dagger$ as long as $a\psi \neq 0$. On the other hand, starting with an eigenfunction of $aa^\dagger$ gives (A.13). For the continuous spectra one can go over to the resolvents of the operators and use commutation relations *à la* Deift. We note that zero modes play an essential role in index problems and in breaking supersymmetry.

The spectrum and the eigensolutions are easily obtained. Let $|0\rangle$ be the ground-state wave function for one oscillator $a|0\rangle = 0$. The ground state of H is given by $|0, 0\rangle := |0\rangle \otimes \begin{pmatrix} 0 \\ 1 \end{pmatrix}$ and fulfils $a|0, 0\rangle = c|0, 0\rangle = 0$. All eigensolutions of H are given by

$$|n, m\rangle = \frac{(a^\dagger)^n}{\sqrt{n!}}\,(c^\dagger)^m|0, 0\rangle, \qquad n = 0, 1, 2, \ldots, \quad m = 0, 1. \qquad (A.14)$$

There is a pure-point spectrum and all excited states are doubly-degenerate. The ground state is not degenerate. We observe that

$$Q|n, 0\rangle \propto |n - 1, 1\rangle, \qquad Q^\dagger|n, 1\rangle \propto |n + 1, 0\rangle. \qquad (A.15)$$

Q maps from 'bosonic' states $|n, 0\rangle$ to 'fermionic' ones $|n - 1, 1\rangle$ with the same energy. $Q^\dagger$ gives the inverse mapping. We may introduce an operator $N_F = \begin{pmatrix} 1 & 0 \\ 0 & 0 \end{pmatrix}$ which counts the number of fermions. The Klein operator $K = (-1)^{N_F}$ gives eigenvalue one for the lower component and -1 for the upper component wave function. We therefore obtain a Z_2 grading of states. A Z_2 grading for operators is obtained by defining even operators to commute with K while odd ones anticommute with K. Q is an odd operator. It generates a symmetry,

$$\delta_Q a^\dagger := [Q, a^\dagger] = \sqrt{\omega}\,c^\dagger, \qquad \delta_Q c := \{Q, c\} = \sqrt{\omega}\,a, \qquad (A.16)$$

and maps from bosons to fermions and vice versa. We have to take the anticommutator in (A.16) between two odd operators Q and c.

After these short introductory remarks on supersymmetric quantum mechanics we cite the generalization which allows the inclusion of general potential interactions in one-dimensional problems (or for the half-line problems). The $N = 2$ superalgebra

$$Q^2 = Q^{\dagger 2} = 0, \qquad \{Q, Q^\dagger\} = H, \qquad [H, Q] = [H, Q^\dagger] = 0 \qquad \text{(A.17)}$$

can be realized in terms of $A = (d/dx) + W(x)$ by putting $Q = \sigma^+ A$ and $Q^\dagger = \sigma^- A^\dagger$. We obtain

$$H = -\frac{d^2}{dx^2} + W^2(x) + \sigma_3 W'(x). \qquad \text{(A.18)}$$

This Hamiltonian consists of a pair of Schrödinger operators which are 'essentially' isospectral. The eigenfunctions of $H_\pm = p^2 + V_\pm(x)$ with $V_\pm = W^2(x) \pm W'(x)$ to non-zero energies are doubly-degenerate. In addition, we observe that the spectrum of H is non-negative. This factorization was used (actually reinvented) by the present authors in order to map one problem with angular momentum ℓ and potential V to another one with angular momentum $\ell + 1$ and potential $\tilde{V}$ (see Eqs. (2.25) and (2.27)).

If $E = \inf \operatorname{spec} H$ equals zero, we call the supersymmetry unbroken. This means that $Q|0\rangle = 0$ and $H|0\rangle = 0$, where $|0\rangle$ denotes the ground-state wave function. Q annihilates the ground state and the symmetry is realized. Broken supersymmetry means $E > 0$.

The algebraic structure mentioned above has its roots in questions which were studied in the eighteenth and nineteenth centuries. If we start from the differential equation $\Phi_0'' = v(x)\Phi_0$ for continuous $v(x)$ and assume that Φ_0 is nowhere vanishing, we may introduce $W(x) = (d/dx) \ln \Phi_0(x)$ and obtain the Ricatti equation (1724) for $W(x)$: $W' + W^2 = v$. Bernoulli had already in 1702 solved the non-linear equation $w' + w^2 + x^2 = 0$ by transforming it to the second-order equation $y'' + x^2 y = 0$.

In 1827 Cauchy asked whether a differential operator could be factorized into a product of first-order differential operators in analogy to the fundamental theorem of algebra. Jacobi proved that any positive self-adjoint differential operator L of order $2m$ can be written as a product of a differential operator p of order m times its adjoint $L = p\bar{p}$. Cayley (1868) transformed $\Phi_0'' = x^{2q-2}\Phi_0$ into $y' + y^2 = x^{2q-2}$. Frobenius first proved the complete factorization which implies the existence of $v_i(x)$ such that

$$L(y) := y^{(n)} + p_1(x)y^{(n-1)} + \ldots = \prod_{i=1}^{n} \left(\frac{d}{dx} - \frac{v_i'(x)}{v_i(x)} \right) y. \qquad \text{(A.19)}$$

Darboux's theorem dates back to 1882 and starts from

$$L\Phi_0 = \left(-\frac{d^2}{dx^2} + v(x) \right) \Phi_0 = 0.$$

L can be written as $L = A^\dagger A$ with $A = (d/dx) - (\ln \Phi_0)'$ if Φ_0 is nowhere vanishing. It asserts that

$$\tilde{L} = AA^+ = -\frac{d^2}{dx^2} + v(x) - 2\frac{d}{dx}\left(\frac{\Phi_0'}{\Phi_0}\right) \tag{A.20}$$

is a solution of $\tilde{L}(A\Phi) = 0$. This result is related to the transformations of Crum and Krein and was used in the inverse scattering problem. For more details on supersymmetric quantum mechanics see, for example, Ref. [154].

Appendix B
Proofs of theorems on angular excitations

We want to give the proofs of the two theorems on the ℓ-dependence of angular excitations:

Theorem B1:
If

$$\frac{d}{dr}\frac{1}{r}\frac{dV}{dr} \gtrless 0, \quad \frac{d^2E}{d\ell^2} \gtrless 0.$$

(B.1)

Then the radial wave function satisfies

$$-\left(\frac{u'}{ru}\right)' - \frac{2(\ell+1)}{r^3} \gtrless 0.$$

(B.2)

Theorem B2:
If

$$\frac{d}{dr}r^2\frac{dV}{dr} \gtrless 0, \quad \frac{d^2E}{d\ell^2} + \frac{3}{\ell+1}\frac{dE}{d\ell} \gtrless 0.$$

(B.3)

Then the radial wave function satisfies

$$-\left(\frac{u'}{u}\right)' - \frac{\ell+1}{r^2} \gtrless 0.$$

(B.4)

Although the beginning of the proofs is the same, we could not reduce them to a single technique.

Naturally one starts from

$$\frac{dE}{d\ell} = (2\ell+1)\int\frac{u^2}{r^2}dr$$

(B.5)

and if we call $v = \partial u/\partial \ell$

$$\frac{d^2E}{d\ell^2} = 2\int\frac{u^2}{r^2}dr + 2(2\ell+1)\int\frac{uv}{r^2}dr$$

(B.6)

(we omit the subscript ℓ).

151

Notice that

$$\int_0^\infty uv\,dr = 0 \tag{B.7}$$

and that, from the Wronskian relation

$$vu' - u'v = \int_0^r \left(\frac{2\ell + 1}{s^2} - \frac{dE}{d\ell} \right) u^2 ds \,, \tag{B.8}$$

v/u is an increasing function. Hence v is negative for small r and positive for large r, and because of (B.7)

$$\int_0^r uv\,ds > 0 \,. \tag{B.9}$$

Beyond this point we have to separate the proofs.

Proof of Theorem B1:
We must bound the integral $\int (uv/r^2)dr$ in (B.6). We shall treat the case of

$$(d/dr)(1/r)(dV/dr) < 0 \,.$$

We write, integrating by parts,

$$\int \frac{uv}{r^2} dr = \int \frac{2dr}{r^3} \int_0^r uv\,dr \,. \tag{B.10}$$

Now from (B.2) and (B.7)

$$\int \frac{uv}{r^2} dr < -\frac{1}{(\ell+1)} \int_0^\infty \left(\frac{u'}{ru} \right)' \int_0^r uv\,ds \tag{B.11}$$

and integrations, again by parts,

$$\int \frac{uv}{r^2} dr < \frac{1}{(\ell+1)} \int_0^\infty \frac{u'v\,dr}{r} \,.$$

Combining

$$\int_0^\infty \left(\frac{u'v + uv'}{r} - \frac{uv}{r^2} \right) dr = 0$$

and, from (B.8),

$$\int_0^\infty \frac{u'v - uv'}{r} - \int \frac{dr}{r} \int_0^r \left(\frac{dE}{d\ell} - \frac{2\ell+1}{s^2} \right) u^2 ds$$

$$= -\int dr \left[(2\ell+1)\frac{\log r}{r^2} - \log r \frac{dE}{dr} \right] u^2 \,,$$

we get

$$\int \frac{uv}{r^2} dr < \frac{1}{2(\ell+1)} \left[\int \frac{uv}{r^2} dr + \int \left(\frac{dr \log r}{r^2}(2\ell+1) - \log r \frac{dE}{d\ell} \right) u^2 \right] \,,$$

and hence, using $\log r = \lim_{\epsilon \to 0}(r^\epsilon - 1/\epsilon)$,

$$(2\ell+1)\int \frac{uv}{r^2}dr < (2\ell+1)\lim_{\epsilon \to 0}\frac{1}{\epsilon}\left[\int u^2 \int \frac{u^2 r^\epsilon}{r^2} - \int \frac{u^2}{r^2}\int u^2 r^\epsilon\right].$$

From the logarithmic convexity of

$$\frac{\int u^2 r^v}{\Gamma\left((2\ell+3+v)/2\right)}$$

we get

$$\int \frac{uv}{r^2}dr < -\frac{1}{2\ell+1}\int \frac{u^2}{r^2}dr$$

and, inserting in (B.6), we get

$$\frac{d^2 E}{d\ell^2} < 0.$$

Clearly, the proof for the case

$$\frac{d}{dr}\frac{1}{r}\frac{dV}{dr} > 0$$

is completely parallel and Theorem B1 is completely established.

Proof of Theorem B2:
This necessitates, as we mentioned in the text, the use of the Chebyshev inequality (2.156):

$$\int f\,h\,dx \int g\,h\,dx \le \int fg\,h dx \int h\,dx,$$

if h is non-negative and f and g are both non-decreasing or both non-increasing.
The proof of this inequality is trivial: the quantity

$$\int dx\,dy\,h(x)h(y)(f(x)-f(y))(g(x)-g(y))$$

is obviously non-negative under the assumptions.
We take the case

$$\frac{d}{dr}r^2\frac{dV}{dr} < 0.$$

Then we calculate $\int(uv/r^2)dr$ in another way:

$$\int \frac{uv}{r^2} = \int \frac{1}{r}(uv)'dr = 2\int_0^\infty \frac{u'v}{r}dr - \int_0^\infty \log r\left(\frac{2\ell+1}{r^2} - \frac{dE}{d\ell}\right)u_\ell^2 dr,$$

$$(B.12)$$

where the Wronskian (B.8) and an integration by parts have been used. On the other hand, the Chebyshev inequality with $f = v/u$, $g = u'/u - (\ell + 1)/r$, $h = u^2/r$ gives, since f and g are increasing,

$$\int \frac{uv}{r} dr \int \left(\frac{1}{r} uu' - \frac{(\ell + 1)}{r^2} u^2 \right) dr < \int \frac{u^2 dr}{r} \int \left(\frac{1}{r} u'v - \frac{\ell + 1}{r^2} uv \right) .$$

(B.13)

Combining (B.12) and (B.13), the integral $\int (u'v/r) dr$ can be eliminated; what remains to be done, to get an inequality on $\int uv dr/r^2$, is to manipulate $\int (uv/r) dr$. For this we do something analogous to (B.9) and (B.10), except that now we have

$$-\left(\frac{u'}{u} \right)' - \frac{\ell + 1}{r^2} < 0$$

so

$$\int \frac{uv dr}{r} < -\frac{1}{\ell + 1} \int_0^\infty \frac{u'}{u} dr \int_0^r uv \, ds$$

$$= \frac{1}{2\ell + 2} \int (u'v - vu') dr$$

$$= \frac{1}{2\ell + 2} \int \frac{2\ell + 1}{r} u^2 dr - \frac{1}{2\ell + 2} \frac{dE}{d\ell} \int ru^2 dr .$$

In the end, we get an inequality on $\int (uv/r^2) dr$ containing only moments of u^2 (including logarithmic ones), for which the standard concavity inequalities can be used. In the end, all inequalities miraculously go in the same direction and the theorem is proved.

Remark:

A simple proof of Theorem B1 can be obtained [48] by using the sum rule (2.19):

$$\frac{1}{2}(E_{\ell+1} - E_\ell)^2 \int r \, u_\ell \, u_{\ell+1} dr = \int \frac{dV}{dr} u_\ell \, u_{\ell+1} dr ,$$

in combination with the fact that $u_{\ell+1}/u_\ell$ is an increasing function of r. However, this method of proof does not seem to work for Theorem B2.

Appendix C
The Sobolev inequality

Here we would like to review our study of the functional

$$F_q(\psi) = \frac{\int d^3x |\nabla \psi|^2}{\left(\int d^3x r^{q-3} |\psi|^{2q}\right)^{1/q}}, \tag{C.1}$$

which led us to obtain a family of optimal conditions for the absence of bound states in a potential [130]. F_q of (C.1) remains unchanged under scale transformations of the form $\psi(x) \to \lambda\psi(\rho x)$ with $\lambda, \rho \neq 0$. Let D be the space of infinitely differentiable functions with compact support. There is no loss of generality if we assume that these functions are real-valued. We intend to determine the numbers

$$\mu_q = \inf_{0 \neq \psi \in D} F_q(\psi). \tag{C.2}$$

It will turn out that μ_q is strictly positive for $1 < q \leq 3$. For the energy functional

$$H(\psi) = \int d^3x |\nabla \psi|^2 - \int d^3x\, V(x)|\psi(x)|^2 \tag{C.3}$$

we can than use (C.1) for the kinetic energy and Hölder's inequality for the potential energy part, obtaining a lower bound of the form

$$H(\psi) \geq \{\mu_q - N_p(V, y)\} \|r^{(q-3)/2q}\psi\|_{2q}^2$$
$$N_p^p(V, y) = \int d^3x |\check{y} - \check{x}|^{2p-3} V^p(\check{x}) \tag{C.4}$$

with $p^{-1} + q^{-1} = 1$. (C.4) is the starting point of our method for obtaining conditions for the absence of bound states in a potential problem.

Since the functional F_q is invariant under rotations around the origin, we might expect that the infimum in (C.2) is to be sought among centrally symmetric functions $\psi(r)$. It turns out that the minimization of the functional F_q^R (being the restriction of F_q to centrally symmetric ψs) is

155

a relatively simple task. The numbers $\mu_q^R = \inf F_q^R(\psi)$ can be explicitly computed and turn out to be strictly positive for $1 \leq q \leq \infty$ (see below).

This naïve argument is, however, wrong in the case $q > 3$; although $\mu_q^R > 0$, we have $\mu_q = 0$ for $q > 3$. For, suppose the contrary. Take a potential of compact support deep enough to bind a particle. Then, since $2p - 3 < 0$ in (C.4), N_p can be made as small as we like by taking $|y|$ big enough, such that $H(\psi) \geq 0$. This contradicts the fact that there is a negative bound state and therefore $\mu_q = 0$ for $q > 3$.

On the other hand, we have the following proposition.

Proposition:
For

$$1 < q \leq 3 : \inf F_q = \inf F_q^R . \tag{C.5}$$

For the proof we remark that we can restrict ourselves to non-negative ψs, because replacing ψ by $|\psi|$ will not change F_q. Next, we use the following theorem.

Rearrangement theorem:
Given $\psi(x) \geq 0$, define $\psi_R(|\check{x}|)$, the spherically decreasing rearrangement of $\psi : \psi_R$ is a decreasing function of $|\check{x}| = r$, such that for every non-negative constant M, the Lebesgue measure $\mu(\psi_R(|\check{x}|) \geq M) = \mu(\psi(\check{x})) \geq M)$. Then

$$\int d^3x |\nabla \psi|^2 \geq \int d^3x |\nabla \psi_R|^2 \tag{C.6}$$

and

$$\int d^3x \psi \chi \leq \int d^3x \, \psi_R \, \chi_R , \tag{C.7}$$

where ψ and χ are any two positive functions.

Now we take $\chi = r^{q-3}$; for $q \leq 3$, χ is decreasing and $\chi_R = \chi$. We have also evidently that $\psi_R^{2q} = (\psi_R)^{2q}$ so that $F_q(\psi) \geq F_q(\psi_R)$. This is just the statement (C.5).

For the spherically symmetric functional F_q^R we have the following theorem.

Theorem:
For $1 < q < \infty$, the functional F_q^R has the strictly positive infimum

$$\mu_q^R = \frac{p}{p-1} \cdot \left(4\pi \frac{(p-1)\Gamma^2(p)}{\Gamma(2p)} \right)^{1/p} , \quad p^{-1} + q^{-1} = 1 , \tag{C.8}$$

which is attained by the uniquely determined family of functions

$$\psi_q(r) = \frac{a}{\left(1 + b\, r^{\frac{1}{p-1}}\right)^{p-1}}, \qquad (C.9)$$

where the arbitrary constants a and b reflect the scale invariance of the problem.

Since the detailed proof is a little delicate [130], we prefer to present here only the formal calculations leading to (C.8) and (C.9). By the change of variables $r \to x = \ln r$ and $\psi \to \phi = \sqrt{r}\psi$, the functional F_q^R takes the form

$$\frac{F_q^R}{(4\pi)^{1/p}} = \frac{\int_{-\infty}^{\infty} dx \left\{ (d\phi/dx)^2 + (1/4)\phi^2 \right\}}{\left(\int_{-\infty}^{\infty} dx\, \phi^{2q} \right)^{1/q}} = \frac{I}{J^{1/q}}. \qquad (C.10)$$

The naïve variational equation $\delta F_q^R = 0$ gives us the differential equation

$$\phi'' = \frac{1}{4}\phi - \kappa\phi^{2q-1}, \;\; \kappa = \frac{I}{J}, \;\; 2 < 2q < \infty, \qquad (C.11)$$

which we have to solve under the initial conditions

$$\phi(\pm\infty) = \phi'(\pm\infty) = 0, \qquad (C.12)$$

since the integrals I and J have to converge. The first integral of (C.11) is given by

$$\phi'^2 = \frac{1}{4}\phi^2 - \frac{\kappa}{q}\phi^{2q}. \qquad (C.13)$$

The arbitrary additive constant on the r.h.s. of (C.13) has been set equal to zero in accordance with (C.12). Up to a translation, the solutions to (C.13) are given by inversion of the integral

$$\int_\phi^1 \frac{2dt}{t\sqrt{1 - 8\kappa t^{2q-2}/q}} = |\tilde{x}|, \;\; 1 < q < \infty, \qquad (C.14)$$

which is an elementary integral with the result

$$\phi_q(x) = \frac{\text{const}}{(\cosh{(q-1)x/2})^{1/(q-1)}}, \qquad (C.15)$$

which is precisely formula (C.9) in the old variables. It remains to compute the minimal values $F_q^R(\phi_q)$. We insert (C.15) into (C.10) and obtain elementary integrals, which can be expressed in terms of Γ-functions, and lead to formula (C.8).

Let us end this appendix by pointing out an amusing fact, which illustrates the necessity of a detailed proof of this theorem. Let F_q^a be the restriction of F_q^R to functions, which vanish outside and on the boundary of a sphere of finite radius a. Then it can be shown that the infimum of F_q^a is identical to that of F_q^R, although there is no function which saturates that minimum.

References

[1] R. Jost, *Helv. Phys. Acta* **19** (1946) 246.

[2] R. Jost and A. Pais, *Phys. Rev.* **82** (1951) 84.

[3] V. Bargmann, *Proc. Acad. Sci. USA* **38** (1952) 961.

[4] J. Schwinger, *Proc. Acad. Sci. USA* **47** (1961) 122.

[5] F.J. Dyson and A. Lenard, *J. Math. Phys.* **8** (1967) 423.

[6] E.H. Lieb and W. Thirring, *Phys. Rev. Lett.* **35** (1975) 687.

[7] A. Martin, *Commun. Math. Phys.* **129** (1990) 161;
B. Helffer and J. Robert, *Asymptotic Analysis* **3** (1990) 91;
B. Helffer and J. Robert, *Ann. Inst. Henri Poincaré Phys. Théor.* **53** (1990) 139;
B. Helffer and B. Parise, preprint Dept. of Mathematics and Computer Sciences, Ecole Normale Supérieure, Paris, 15.1.1990.

[8] J.M. Levy Leblond, *J. Math. Phys.* **10** (1969) 806.

[9] J.L. Basdevant, A. Martin and J.-M. Richard, *Nucl. Phys.* **B340** (1990) 60, 69.

[10] R. Vinh Mau, in *Mesons in Nuclei,* vol. 1 (eds. M. Rho and D. Wilkinson) p. 151, Amsterdam: North Holland (1979).

[11] G.F. Chew, *The Analytic S Matrix, a Basis for Nuclear Democracy,* New York: Benjamin (1966).

[12] M. Gell-Mann, in *Proceedings of the 1962 International Conference on High Energy Physics,* CERN (ed. J. Prentki) p. 805, Geneva: CERN (1962).

[13] R.H. Dalitz, in *Oxford International Conference on Elementary Particles,* 1965 (ed. T.R. Walsh) p. 157, Chilton: Rutherford High Energy Laboratory (1966).

[14] S.B. Gerasimov, *Zh. Eksper. Teor. Fiz. (USSR)* **50** (1966) 1559, English translation *Sov. Phys. JETP* **23** (1966) 1040.

[15] J.J. Aubert *et al., Phys. Rev. Lett.* **33** (1974) 1404.

[16] J.E. Augustin *et al.*, *Phys. Rev. Lett.* **33** (1974) 1406.

[17] G.S. Abrams *et al.*, *Phys. Rev. Lett.* **33** (1974) 1453.

[18] E. Eichten *et al.*, *Phys. Rev. Lett.* **34** (1975) 369.

[19] T. Applequist *et al.*, *Phys. Rev. Lett.* **34** (1975) 365.

[20] A. De Rujula, H. Georgi and S.L. Glashow, *Phys. Rev.* **D12** (1975) 147.

[21] Y.B. Zeldovitch and A.D. Sakharov, *Acta Phys. Hung.* **22** (1967) 153;
A.D. Sakharov, *Zh. Eksper. Teor. Fiz. (USSR)* **78** (1980) 2112;
English translation *Sov. Phys. JETP* **51** (1980) 1059.

[22] P. Federman, H.R. Rubinstein and I. Talmi, *Phys. Lett.* **22** (1966) 208.

[23] D.P. Stanley and D. Robson, *Phys. Rev.* **D21** (1980) 3180.

[24] D.P. Stanley and D. Robson, *Phys. Rev. Lett.* **45** (1980) 235.

[25] N. Isgur, in *Proceedings of the International Conference on High Energy Physics*, Madison, 1980 (eds. L. Durand and L.G. Pondrom) (AIP Conference Proceedings, no. 68) p. 30, New York: AIP (1981);
See also N. Isgur and P. Kociuk, *Phys. Rev.* **D21** (1980) 1868.

[26] J.-M. Richard and P. Taxil, *Phys. Lett.* **B123** (1983) 4531;
Ann. Phys. (New York) **150** (1983) 267.

[27] S. Ono and F. Schöberl, *Phys. Lett.* **B118** (1982) 419.

[28] J.L. Basdevant and S. Boukraa, *Ann. Phys. (Fr.)* **10** (1985) 475; *Z. Phys.* **C30** (1986) 103.

[29] DO Collaboration, *Proceedings of the 27th International Conference on High Energy Physics*, Glasgow, 1994; P.J. Bussey and J.G. Knowles, eds. Institute of Physics, Bristol and Philadelphia (1995).

[30] P. Darriulat, Conference Summary, *27th International Conference on High Energy Physics*, Glasgow, 1994, ibid. p. 367.

[31] A. Martin, in *Heavy Flavours, Nucl. Phys. (Proc. Suppl.)* **1B** (1988) 133;
In this article, toponium widths should be multiplied by 2 (the fact that toponium contains 2 quarks was overlooked!).

[32] DO and CDF Collaborations, *Proceedings of the Rencontres de Physics de la Vallée d'Aoste*, March 1996, to be published by Editions Frontières (ed. M. Greco).

[33] W. Buchmüller, G. Grunberg and S. H.-H. Tye, *Phys. Rev. Lett.* **45** (1980) 103;
W. Buchmüller and S. H.-H. Tye, *Phys. Rev.* **D24** (1981) 132.

[34] A. Martin, *Phys. Lett.* **B100** (1981) 511.

[35] H. Schröder, in *Les Rencontres de Physique de la Vallée d'Aoste*, La Thuile, 1989 (ed. M. Greco) p. 461, Gif-sur-Yvette: Editions Frontières (1989).

[36] A. Martin, *Phys. Lett.* **B223** (1989) 103.

[37] Y. Cen, in *Proceedings of the 29th Rencontres de Moriond, QCD and High Energy Hadronic Interactions*, Méribel 1994, (ed. J. Tran Thanh Van) p. 497, Gif-sur-Yvette: Editions Frontières (1994).

[38] A. Martin, in *Heavy Flavours and High Energy Collisions in the 1–100 TeV Range* (eds. A. Ali and L. Cifarelli) p. 141, New York: Plenum (1989).

[39] J.D. Bjorken, in *Hadron Spectroscopy* (ed. J. Oneda) p. 390, New York: AIP (1985);
See also A. Martin, in *Heavy Flavours and High Energy Collisions in the 1–100 TeV Range* (eds. A. Ali and L. Cifarelli) p. 141, New York: Plenum (1989).

[40] J.P. Ader, J.-M. Richard and P. Taxil, *Phys. Rev.* **D12** (1982) 2370;
S. Nussinov, *Phys. Rev. Lett.* **51** (1983) 2081;
J.-M. Richard, *Phys. Lett.* **B139** (1984) 408.

[41] J.-M. Richard, *Phys. Lett.* **B100** (1981) 515.

[42] V.E. Barnes *et al.*, *Phys. Rev. Lett.* **12** (1964) 204.

[43] S. Godfrey and N. Isgur, *Phys. Rev.* **D32** (1985) 232;
S. Capstick and N. Isgur, *Phys. Rev.* **D34** (1986) 2809;
see also K.T. Chao, N. Isgur and G. Karl, *Phys. Rev.* **D23** (1981) 155.

[44] J.-M. Richard and P. Taxil, *Phys. Rev. Lett.* **54** (1985) 847;
A. Martin, J.-M. Richard and P. Taxil, *Phys. Lett.* **B176** (1986) 224.

[45] E.H. Lieb, *Phys. Rev. Lett.* **54** (1985) 1987.

[46] E. Seiler, *Phys. Rev.* **D18** (1978) 133.

[47] B. Baumgartner, H. Grosse and A. Martin, *Phys. Lett.* **B146** (1984) 363;
Nucl. Phys. **B254** (1985) 528;
Simplified proofs have been given by M.S. Ashbaugh and R.D. Benguria, *Phys. Lett.* **A131** (1988) 273;
and finally by A. Martin, *Phys. Lett.* **A147** (1990) 1.

[48] A.K. Common and A. Martin, *Europhys. Lett.* **4** (1987) 1349.

[49] A. Martin, J.-M. Richard and P. Taxil, *Nucl. Phys.* **B329** (1990) 327.

[50] A. Martin and J. Stubbe, *Europhys. Lett.* **14** (1991) 287;
Nucl. Phys. **B367** (1991) 158.

[51] R. Engfer *et al.*, *At. Data Nucl. Data Tables* **14** (1974) 509.

[52] These figures are taken from E. Chpolsky, *Physique Atomique*, Moscow: Mir, French translation (1978).

[53] S. Bashkin and J.O. Stoner Jr, *Tables of Atomic Energy Levels and Grotrian Diagrams*, vols. I, II, Amsterdam: North-Holland (1978).

[54] J. Stubbe and A. Martin, *Phys. Lett.* **B271** (1991) 208.

[55] W. Kwong and J.L. Rosner, *Phys. Rev.* **D38** (1988) 279.

[56] H. Grosse, A. Martin and J. Stubbe, *Phys. Lett.* **B255** (1991) 563.

[57] H. Grosse, *Phys. Lett.* **B197** (1987) 413.

[58] H. Grosse, A. Martin and J. Stubbe, *Phys. Lett.* **B284** (1992) 347.

[59] C. Quigg and J.L. Rosner, *Phys. Rep.* **56** (1979) 167.

[60] H. Grosse and A. Martin, *Phys. Rep.* **60** (1980) 341.

[61] A. Martin, in *Recent Developments in Mathematical Physics, Proceedings of the XXVI Universitätswochen für Kernphysik*, Schladming, 1987 (eds. H. Mitter and L. Pittner) p. 53, Berlin: Springer (1987).

[62] A. Martin, in *Present and Future of Collider Physics* (eds. C. Bacci et al.) p. 339, Bologna: Italian Physical Society (1990).

[63] A. Martin, in *From Superstrings to the real Superworld, Proceedings of the International School of Subnuclear Physics* (ed. A. Zichichi) p. 482, Singapore: World Scientific (1993).

[64] W. Lucha and F. Schöberl, *Phys. Rep.* **200** (1991) 127.

[65] P. Falkensteiner, H. Grosse, F. Schöberl and P. Hertel, *Comput. Phys. Commun.* **34** (1985) 287.

[66] K. Chadan and M.L. Bateman, *Ann. I.H.P.*, Vol. XXIV No. 1 (1975) 1.

[67] W. Thirring, *A Course on Mathematical Physics III, Quantum Mechanics of Atoms and Molecules*, New York: Springer (1981).

[68] J.D. Bessis, private communication.

[69] G. Feldman, T. Fulton and A. Devoto, *Nucl. Phys.* **B154** (1979) 441.

[70] H. Grosse and A. Martin, *Phys. Lett.* **B134** (1984) 368.

[71] B. Baumgartner, H. Grosse and A. Martin, *Phys. Lett.* **B146** (1984) 363.

[72] M.S. Ashbaugh and R.D. Benguria, *Phys. Lett.* **A131** (1988) 273.

[73] A. Martin, *Phys. Lett.* **A147** (1990) 1.

[74] H. Grosse, A. Martin and A. Pflug, *C. R. Acad. Sci. Paris* **299** série II (1984) 5.

[75] E.U. Condon and G.H. Shortley, *The Theory of Atomic Spectra*, Cambridge: Cambridge University Press (1953).

[76] J.A. Wheeler, *Rev. Mod. Phys.* **21** (1949) 133.

[77] B. Baumgartner, H. Grosse and A. Martin, *Fizika* **17** (1985) 279.

[78] B. Baumgartner, H. Grosse and A. Martin, *Nucl. Phys.* **B254** (1985) 528.

[79] K. Chadan and H. Grosse, *J. Phys.* **A16** (1983) 955.

[80] U. Sukhatme and T. Imbo, *Phys. Rev.* **D28** (1983) 418.

[81] J.J. Loeffel, A. Martin, B. Simon and A.S. Wightman, *Phys. Lett.* **B30** (1969) 656.

[82] S. Flügge, *Practical Quantum Mechanics*, vol. 1, New York: Springer (1971).

[83] J.-M. Richard, private communication.

[84] R.M. Sternheimer, in *Progress in Atomic Spectroscopy*, part C (eds. H.J. Beyer and H. Kleinpoppen) New York: Plenum (1984).

[85] A.K. Common and A. Martin, *J. Phys. A: Math. Gen.* **20** (1987) 4.

[86] A.K. Common and A. Martin, *J. Math. Phys.* **30** (1989) 601.

[87] R.E. Prange and S.M. Girvin (eds.), *The Quantum Hall Effect* (Contemporary Physics), New York: Springer (1987).

[88] H. Grosse, A. Martin and J. Stubbe, *Phys. Lett.* **A181** (1993) 7; H. Grosse and J. Stubbe, *Lett. Math. Phys.* **34** (1995) 59.

[89] H.L. Cycon, R.G. Froese, W. Kirsch and B. Simon, *Schrödinger Operators with Applications to Quantum Mechanics and Global Geometry* (Texts and Monographs in Physics) Berlin: Springer (1987).

[90] A.K. Common, *J. Phys. A: Math. Gen.* **18** (1985) 221.

[91] A.K. Common, A. Martin and J. Stubbe, *Commun. Math. Phys.* **134** (1990) 509.

[92] R.S. Conti, S. Hatamian, L. Lapidus, A. Rich and M. Skalsey, *Phys. Lett.* **A177** (1993). For a review prior to this experiment see A.P. Mills in *Quantum Electrodynamics* (ed. T. Kinoshita) p. 774, Singapore: World Scientific (1990).

[93] W. Kwong, J. Rosner and C. Quigg, *Annu. Rev. Nucl. Part. Sci.* **37** (1987) 325.

[94] E 760 Collaboration, T.A. Armstrong *et al.*, *Phys. Rev. Lett.* **69** (1992) 2337; This experiment confirms the signal first observed by C. Baglin *et al.*, *Phys. Lett.* **B171** (1986) 135.

[95] V. Singh, S.N. Biswas and K. Datta, *Phys. Rev.* **D18** (1978) 1901.

[96] A.V. Turbiner, *Commun. Math. Phys.* **118** (1988) 467.

[97] M. Krammer and H. Krasemann, *Acta Phys. Austriaca*, Suppl. XXI (1979) 259.

[98] A. Martin, *Phys. Lett.* **B70** (1977) 194.

[99] Ref. [60], p. 359.

[100] A. Martin, *Phys. Rep.* **134** (1986) 305.

[101] R. Bertlmann and A. Martin, *Nucl. Phys.* **B168** (1980) 111.

[102] A.K. Common, *Nucl. Phys.* **B224** (1983) 229; *Nucl. Phys.* **B184** (1981) 323.

[103] A.K. Common, *Nucl. Phys.* **B162** (1980) 311.

[104] A.K. Common, *J. Math. Phys.* **82** (1991) 3111.

[105] A. Martin and J. Stubbe, *Z. Phys.* **C62** (1994) 167.

[106] B.E. Palladino and P. Leal-Ferreira, *Phys. Lett.* **B185** (1987) 118.

[107] J.S. Kang and H.J. Schnitzer, *Phys. Rev.* **D12** (1975) 841.

[108] H. Grosse, A. Martin and J. Stubbe, *Phys. Lett.* **B284** (1992) 347.

[109] C.V. Sukumar, *J. Phys.* **A18** (1985) L 697.

[110] C. Itzykson and J.B. Zuber, *Quantum Field Theory*, New York: McGraw-Hill (1985).

[111] M.E. Rose, *Relativistic Electron Theory*, New York: Wiley (1961).

[112] R.L. Kelly, *J. Phys. Chem. Ref. Data* **16** (1987) Suppl. 1.

[113] K. Chadan and P.C. Sabatier, *Inverse Problems in Quantum Scattering Theory*, 2nd ed. (Springer Texts and Monographs in Physics) New York: Springer (1989).

[114] H.B. Thacker, C. Quigg and J.L. Rosner, *Phys. Rev.* **D18** (1978) 274, 288.

[115] H. Grosse and A. Martin, *Nucl. Phys.* **B148** (1979) 413.

[116] J.F. Schonfeld, W. Kwong, J.L. Rosner, C. Quigg and H.B. Thacker, *Ann. Phys. (New York)* **128** (1980) 1.

[117] W. Kwong and J.L. Rosner, *Phys. Rev.* **D38** (1988) 273.

[118] G.M. Gasymov and B.M. Levitan, *Russ. Math. Surv.* **19** (1964) 1.

[119] R. Jost and W. Kohn, *Phys. Rev.* **88** (1952) 382.

[120] R.P. Boas, *Entire Functions*, New York: Academic Press (1954).

[121] A. Martin, *Helv. Phys. Acta* **45** (1972) 140.

[122] G. Szegö, *Orthogonal Polynomials*, p. 108, (A.M.S. colloquium publ. vol. 23), Providence, R.I.: Am. Math. Soc. (1975).

[123] N.K. Nikol'skij, *Functional Analysis I*, Berlin: Springer, (1992).

[124] A. Schönhagen, *Approximationstheorie*, p. 49, Berlin: W. de Gruyter (1971).

[125] T. Regge, *Nuovo Cimento* **14** (1959) 951.

[126] J.J. Loeffel and A. Martin, *Publication de la RCP* No. 25, Vol. 11 (Département de Mathématiques, Université de Strasbourg, 1970).

[127] E.H. Lieb and W.E. Thirring, in *Studies in Mathematical Physics* (eds. E.H. Lieb, B. Simon and A.S. Wightman) p. 269, Princeton University Press, Princeton: New Jersey (1976).

[128] K. Chadan, *Nuovo Cimento* **A58** (1968) 191.

[129] A. Martin, *Helv. Phys. Acta* **45** (1972) 142.

[130] V. Glaser, A. Martin, H. Grosse and W. Thirring, in *Studies in Mathematical Physics* (eds. E.H. Lieb, B. Simon and A.S. Wightman) p. 169, Princeton University Press, Princeton: New Jersey (1976).

[131] A. Martin, *Commun. Math. Phys.* **69** (1979) 89; **73** (1980) 79.

[132] Ph. Blanchard and J. Stubbe, *Rev. Math. Phys.* **8** (1996) 503.

[133] F. Calogero, *Commun. Math. Phys.* **1** (1965) 80; J.H.E. Cohn, *J. London Math. Soc.* **40** (1965) 523.

[134] E.H. Lieb, *Bull. Am. Math. Soc.* **82** (1976) 751.

[135] V. Glaser, H. Grosse and A. Martin, *Commun. Math. Phys.* **59** (1978) 197.

[136] P. Li and S.T. Yau, *Commun. Math. Phys.* **88** (1983) 309.

[137] Ph. Blanchard, J. Stubbe and J. Rezende, *Lett. Math. Phys.* **14** (1987) 215.

[138] C.S. Gardner, J.M. Greene, M.D. Kruskal and R.M. Miura, *Commun. Pure Appl. Math.* **27** (1974) 97.

[139] M. Aizenman and E.H. Lieb, *Phys. Lett.* **66A** (1978) 427.

[140] H. Grosse, *Acta Phys. Austriaca* **52** (1980) 89.

[141] K. Chadan and A. Martin, *Commun. Math. Phys.* **70** (1979) 1.

[142] K. Chadan and A. Martin, *Commun. Math. Phys.* **53** (1977) 221;
K. Chadan and H. Grosse, *J. Phys.* **A 16** (1983) 955.

[143] S. Nussinov, *Z. Phys.* **C3** (1979) 165.

[144] A. Martin, *Phys. Lett.* **B103** (1981) 51, and in *The Search for Beauty, Charm and Truth* (eds. G. Bellini and S.C.C. Ting) p. 501, New York: Plenum (1981).

[145] This unofficial number from Opal and Delphi at LEP happens to agree with the previous indications of the Zichichi group at the ISR and of UA1.

[146] R. Askey, *Trans. Am. Math. Soc.* **179** (1973) 71.

[147] A. Martin, CERN preprint TH 6544/92, unpublished.

[148] A. Martin, *Phys. Lett.* **B287** (1992) 251.

[149] A. Martin and J.-M. Richard, *Phys. Lett.* **B185** (1987) 426.

[150] L. Hall and H. R. Post, *Proc. Roy. Soc.* **90** (1967) 381.

[151] J.L. Basdevant, A. Martin, J.-M. Richard and T.T. Wu, *Nucl. Phys.* **B393** (1993) 111.

[152] A. Gonzalez-Garcia, *Few-Body Systems* **8** (1990) 73; **10** (1991) 73.

[153] J.-M. Richard, *Phys. Rep.* **212** (1992) 1.

[154] H. Grosse, in *Recent Developments in Quantum Mechanics* (eds. A. Boutet de Monvel *et al.*) p. 299, Dordrecht: Kluwer (1991).

Index